Denis Caromel Rodney R. Oldehoeft
Marydell Tholburn (Eds.)

Computing
in Object-Oriented
Parallel Environments

Second International Symposium, ISCOPE 98
Santa Fe, NM, USA, December 8-11, 1998
Proceedings

Springer

Gerhard Goos, Karlsruhe University, Germany
Juris Hartmanis, Cornell University, NY, USA
Jan van Leeuwen, Utrecht University, The Netherlands

Volume Editors

Denis Caromel
University of Nice – INRIA Sophia Antipolis
2004 Route des Lucioles, BP 93, F-06902 Sophia Antipolis Cedex, France
E-mail: caromel@unice.fr

Rodney R. Oldehoeft
Colorado State University, Computer Science Department
Fort Collins, CO 80523, USA
E-mail: rro@cs.colostate.edu

Marydell Tholburn
Los Alamos National Laboratory, CIC-8, MS B272
Los Alamos, NM 87545, USA
E-mail: marydell@lanl.gov

Cataloging-in-Publication data applied for

Die Deutsche Bibliothek - CIP-Einheitsaufnahme

Computing in object oriented parallel environments : second
international symposium ; proceedings / ISCOPE 98, Santa Fe, NM,
December 8 - 11, 1998. Denis Caromel ... (ed.). - Berlin ; Heidelberg
; New York ; Barcelona ; Hong Kong ; London ; Milan ; Paris ;
Singapore ; Tokyo : Springer, 1998
　(Lecture notes in computer science ; Vol. 1505)
　ISBN 3-540-65387-2

CR Subject Classification (1998): D, G.1-2,F.3

ISSN 0302-9743
ISBN 3-540-65387-2 Springer-Verlag Berlin Heidelberg New York

Typesetting: Camera-ready by author
SPIN 10638994   06/3142 – 5 4 3 2 1 0   Printed on acid-free paper

Preface

This volume contains the Proceedings of the International Symposium on Computing in Object-Oriented Parallel Environments (ISCOPE '98), held at Santa Fe, New Mexico, USA on December 8–11, 1998. ISCOPE is in its second year,[1] and continues to grow both in attendance and in the diversity of the subjects covered. ISCOPE'97 and its predecessor conferences focused more narrowly on scientific computing in the high-performance arena. ISCOPE '98 retains this emphasis, but has broadened to include discrete-event simulation, mobile computing, and web-based metacomputing.

The ISCOPE '98 Program Committee received 39 submissions, and accepted 10 (26%) as Regular Papers, based on their excellent content, maturity of development, and likelihood for widespread interest. These 10 are divided into three technical categories.

Applications: The first paper describes an approach to simulating advanced nuclear power reactor designs that incorporates multiple local solution methods and a natural extension to parallel execution. The second paper discusses a Time Warp simulation kernel that is highly configurable and portable. The third gives an account of the development of software for simulating high-intensity charged particle beams in linear particle accelerators, based on the POOMA framework, that shows performance considerably better than an HPF version, along with good parallel speedup.

Runtime and Libraries: The first paper in this category evaluates Java as a language and system for high-performance numerical computing, exposing some issues to face in language features and compilation strategies. The second describes using the Illinois Concert system to parallelize an adaptive mesh refinement code, showing that a combination of aggressive compiler optimizations and advanced run-time support can yield good parallel performance for dynamic applications. The third paper presents a unified framework for building a numerical linear algebra library for dense and sparse matrices, achieving high performance and minimizing architectural dependencies. In the fourth paper, a parallel run-time substrate is presented that supports a global addressing scheme, object mobility, and automatic message forwarding for implementing adaptive applications on distributed-memory machines.

Numerics and Algorithms: The first paper describes a software package for partitioning data on structured grids, supporting inherent and new partitioning algorithms, and describes its use in two applications. The second describes a family of multiple minimum degree algorithms to generate permutations of large, sparse, symmetric matrices to minimize time and space required in matrix factorization. The final regular paper discusses two optimizing transformations for numerical frameworks, one that reduces inter-processor communication and another that improves cache utilization.

In addition, the Program Committee selected 15 submissions as Short Papers. These papers were deemed to represent important work of a more specialized

---

[1] The ISCOPE'97 Proceedings are available from Springer as LNCS, Vol. 1343.

nature or to describe projects that are still in development. The Short Papers are divided into four technical categories.

Metacomputing: The first paper presents a Java-based infrastructure to combine web-based metacomputing with cluster-based parallel computing. The second describes an experimental metacomputing system that is dynamically reconfigurable in its use of systems and networks, and also in its own capabilities. The third paper outlines a distributed platform to ease the combination of heterogeneous networks, concentrating on the design of its kernel software. The fourth paper presents language constructs for the simultaneous creation of entire static object networks which have useful properties.

Frameworks and Run-time: The first paper describes a class library for FIFO queues that can be incorporated with Time Warp simulation mechanisms and retain the advantages of inlined data structures and efficient state saving. The second paper presents a thread profiling system that is cognizant of the underlying concurrent run-time environment. The third paper evaluates a high-level, portable, multithreaded run-time system for supporting concurrent object-oriented languages. The fourth describes a run-time library for data-parallel applications that covers a spectrum of parallel granularities, problem regularities and user-defined data structures. The last paper in this section describes the design and use of a component architecture for large-scale simulations of scientific problems, based in turn on the POOMA framework.

Numerics and Algorithms: The first paper discusses the parallelization and implementation of Monte Carlo simulations for physical problems. The second presents a parallel implementation of the dynamic recursion method for tridiagonalizing sparse matrices efficiently. The third discusses the design of software for solving sparse, symmetric systems of linear equations by direct methods. The fourth paper describes a template library of two-phase container classes and communication primitives for parallel dynamic mesh applications.

Arrays: The first of two papers in this category describes the Blitz++ library, meant to provide a base environment of vectors, arrays and matrices for scientific computing with C++. The second discusses the design of arrays and expression evaluation strategies in the new POOMA II framework development.

This collection of 25 papers represents the state of the art in applying object-oriented methods to parallel computing. ISCOPE '98 is truly international in scope, with its 72 contributing authors representing 24 research institutions in 9 countries. The ISCOPE '98 organizers are confident that the reader will share their excitement about this dynamic and important area of computer science and applications research.

At the end of this volume, the Author Contacts section details the affiliations, postal addresses, and email addresses of all the proceedings authors.

ISCOPE '98 is partially supported by the Mathematical, Information, and Computational Sciences Division, Office of Energy Research, U.S. Department of Energy.

Steering Committee

Dennis Gannon, Indiana University
Denis Caromel, University of Nice–INRIA Sophia Antipolis
Yutaka Ishikawa, Real World Computing Partnership
John Reynders, Los Alamos National Lab
Satoshi Matsuoka, Tokyo Institute of Technology
Jörg Nolte, German National Research Center for Information Technology

Organizing Chairs

Dennis Gannon, Indiana University, General Chair
Denis Caromel, University of Nice–INRIA Sophia Antipolis, Program
Yutaka Ishikawa, Real World Computing Partnership, Posters
John Reynders, Los Alamos National Lab, Workshops/BOF
Rodney R. Oldehoeft, Colorado State University, Proceedings
Marydell Tholburn, Los Alamos National Lab, Local Arrangements/Publicity

Program Committee

Ole Agesen, Sun Microsystems Labs, USA
Denis Caromel, University of Nice–INRIA Sophia Antipolis, France
Antonio Corradi, University of Bologna, Italy
Geoffrey Fox, Northeast Parallel Architecture Center, Syracuse Univ., USA
Dennis Gannon, Indiana University, USA
Jean-Marc Geib, University of Lille, France
Andrew Grimshaw, University of Virginia, USA
Urs Hölzle, University of California-Santa Barbara, USA
Yutaka Ishikawa, Real World Computing Partnership, Japan
Jean-Marc Jezequel, IRISA/CNRS, France
Pierre Kuonen, EPFL, Switzerland
Satoshi Matsuoka, Tokyo Institute of Technology, Japan
Jörg Nolte, Institute GMD-FIRST, Germany
Rodney R. Oldehoeft, Colorado State University, USA
John Reynders, Los Alamos National Lab, USA
Wolfgang Schroeder-Preikschat, Magdeburg, GMD, Germany
Anthony Skjellum, Mississippi State University, USA
David F. Snelling, Fujitsu European Center for Information Technology, UK

Kenjiro Taura, University of Tokyo, Japan
MaryDell Tholburn, Los Alamos National Lab, USA
Andrew L. Wendelborn, University of Adelaide, Australia
Russel Winder, King's College London, UK

Additional External Referees

Cliff Addison	Sven van den Berghe	Danilo Beuche
Lars Buettner	P. Calegari	Peter Chow
Paul Coddington	George Crawford	David Detlefs
Rossen Dimitrov	Stephane Ecolivet	Antônio A. Fröhlich
Robert George	Matt Gleeson	Abdelaziz Guerrouat
F. Guidec	Fredéric Guyomarc'h	Ute Haack
Peter Harrison	Greg Henley	Wai Ming Ho
Naoki Kobayashi	Evelina Lamma	Pascale Launay
Andrea Omicini	Jean-Louis Pazat	R. Radhakrishnan
Tomasz Radzik	Y.S. Ramakrishna	C. Stefanelli
F. Zambonelli		

Table of Contents

Regular Papers

Applications

Object-Oriented Approach for an Iterative Calculation Method and Its
Parallelization with Domain Decomposition Method 1
  Masahiro Tatsumi, Akio Yamamoto

An Object-Oriented Time Warp Simulation Kernel 13
  *Radharamanan Radhakrishnan, Dale E. Martin, Malolan Chetlur,
  Dhananjai Madhava Rao, Philip A. Wilsey*

Particle Beam Dynamics Simulations Using the POOMA Framework 25
  *William Humphrey, Robert Ryne, Timothy Cleland, Julian Cummings,
  Salman Habib, Graham Mark, Ji Qiang*

Runtime and Libraries

An Evaluation of Java for Numerical Computing 35
  Brian Blount, Siddhartha Chatterjee

High-Level Parallel Programming of an Adaptive Mesh Application Using
the Illinois Concert System 47
  Bishwaroop Ganguly, Andrew Chien

The Matrix Template Library: A Generic Programming Approach to High
Performance Numerical Linear Algebra 59
  Jeremy G. Siek, Andrew Lumsdaine

The Mobile Object Layer: A Run-Time Substrate for Mobile Adaptive
Computations 71
  Nikos Chrisochoides, Kevin Barker, Démian Nave, Chris Hawblitzel

Numerics and Algorithms I

Software Tools for Partitioning Block-Structured Applications 83
  Jarmo Rantakokko

An Object-Oriented Collection of Minimum Degree Algorithms 95
  Gary Kumfert, Alex Pothen

Optimizing Transformations of Stencil Operations for Parallel Object-
Oriented Scientific Frameworks on Cache-Based Architectures 107
  Federico Bassetti, Kei Davis, Dan Quinlan

Short Papers

Metacomputing

Merging Web-Based with Cluster-Based Computing 119
  Luís Moura Silva, Paulo Martins, João Gabriel Silva

Dynamic Reconfiguration and Virtual Machine Management in the
Harness Metacomputing System 127
  Mauro Migliardi, Jack Dongarra, Al Geist, Vaidy Sunderam

JEM-DOOS: The Java/RMI Based Distributed Objects Operating System
of the JEM Project 135
  Serge Chaumette

Static Networks: A Powerful and Elegant Extension to Concurrent Object-
Oriented Languages 143
  Josh Yelon, Laxmikant V. Kalé

Frameworks and Runtime

A FIFO Queue Class Library as a State Variable of Time Warp Logical
Processes 151
  Soichiro Hidaka, Terumasa Aoki, Hitoshi Aida, Tadao Saito

μProfiler: Profiling User-Level Threads in a Shared-Memory Programming
Environment 159
  Peter A. Buhr, Robert Denda

Evaluating a Multithreaded Runtime System for Concurrent
Object-Oriented Languages 167
  Antonio J. Nebro, Ernesto Pimentel, José M. Troya

Object-Oriented Run-Time Support for Data-Parallel Applications 175
  Hua Bi, Matthias Kessler, Matthias Wilhelmi

Component Architecture of the Tecolote Framework 183
  Mark Zander, John Hall, Jim Painter, Sean O'Rourke

Numerics and Algorithms II

Parallel Object Oriented Monte Carlo Simulations 191
  Matthias Troyer, Beat Ammon, Elmar Heeb

A Parallel, Object-Oriented Implementation of the Dynamic Recursion
Method 199
  Wolfram T. Arnold, Roger Haydock

Object-Oriented Design for Sparse Direct Solvers 207
  Florin Dobrian, Gary Kumfert, Alex Pothen

Janus: A C++ Template Library for Parallel Dynamic Mesh Applications . 215
Jens Gerlach, Mitsuhisa Sato, Yutaka Ishikawa

Arrays

Arrays in Blitz++ 223
Todd L. Veldhuizen

Array Design and Expression Evaluation in POOMA II 231
Steve Karmesin, James Crotinger, Julian Cummings, Scott Haney,
William Humphrey, John Reynders, Stephen Smith, Timothy Williams

Author Contacts 239

Author Index.. 243

Object-Oriented Approach for an Iterative Calculation Method and Its Parallelization with Domain Decomposition Method

Masahiro Tatsumi[1,2]* and Akio Yamamoto[1]

[1] Nuclear Fuel Industries, Ltd., Osaka, Japan
[2] Osaka University, Osaka, Japan

Abstract. With trends toward more complex nuclear reactor designs, advanced methods are required for appropriate reduction of design margins from an economical point of view. As a solution, an algorithm based on an object-oriented approach has been developed. In this algorithm, calculation meshes are represented as calculation objects wherein specific calculation algorithms are encapsulated. Abstracted data, which are neutron current objects, are exchanged between these objects. Calculation objects can retrieve required data having specified data types from the neutron current objects, which leads to a combined use of different calculation methods and algorithms in the same computation. Introducing a mechanism of object archiving and transmission has enabled a natural extension to a parallel algorithm. The parallel solution is identical with the sequential one. The SCOPE code, an actual implementation of our algorithm, showed good performance on a networked PC cluster, for sufficiently coarse granularity.

1 Introduction

The generation of electricity by nuclear energy in Japan supplies approximately 30% of total power generation. Nuclear power is becoming more important from the viewpoint of preventing the greenhouse effect, and of providing sufficient supply against increasing power demand. Advanced degrees of safety are fundamentally required to utilize nuclear power, because of the social impact in case of severe accidents. At the same time, efficient design is also important from an economical point of view.

Good designs can improve efficiency and save energy generation costs. Thus advanced and precise design tools have been become very important. So far, several kinds of approximations[1] in modeling actual reactor cores have been developed and adopted in order to perform calculations using computers with limited performance. While those models have contributed to a reduction in computing time while maintaining required precision, they are not suitable for direct use in much more complex future designs. Therefore, a more comprehensive technique in problem modeling and its application to reactor design will

---

* Also a Ph.D. candidate at the Graduate School of Engineering, Osaka University.

be needed. However, it will not be sufficient simply to perform high-quality calculations with complex designs because of the associated high computation cost. Consequently, an optimization technique for the calculation method itself is needed that results in realistic computation times and high-accuracy solutions when handling actual designs.

We have developed a new calculation technique using an object-oriented approach that enables coupling of arbitrary calculation theories in the same calculation stage. In this technique, an object based on a particular calculation theory is allocated for each calculation mesh in the finite difference method. Those calculation meshes have the same interface for data exchange with adjacent meshes, which enables the coupling of different calculation theories, providing an "adequate theory for any domain."

In the next section, analysis of reactor cores by neutronics calculations is briefly reviewed. Techniques with an object-oriented approach and its natural extension to the parallel algorithm are explained in Sect. 3 and 4, respectively. Several results of a performance analysis are shown in Sect. 5. After some discussion in Sect. 6, we conclude in Sect. 7. Finally plans for further study are shown in Sect. 8.

2 Reactor Core Analysis

In a reactor core, steam is generated by thermal energy produced by controlled fission chain reactions. The steam is injected into turbines that produce electrical energy and is finally returned into the core after condensing to liquid water. A large portion of fission reactions is directly caused by neutrons. Therefore, it is quite important to estimate accurately the space and energy distribution of neutrons in the reactor. However, the size of a neutron is extremely small, compared to that of the reactor, with a wide range of energy distributions. This makes a microscopic analysis by handling neutrons explicitly very difficult. So an approximation is needed.

In the next two subsections, the computation model of reactor core analysis, and solution methods to solve the neutron transport equation in the reactor core are described briefly.

2.1 Computation Model

In a typical analysis technique, three modeling stages are assumed for space and energy in interactions of neutrons and materials: microscopic, mesoscopic and macroscopic stages. Such models are widely used in other kinds of analysis, and boundaries among the three stages may be obscured with improvement in the calculation methods.

Our aim is to perform more precise calculations by integrating the mesoscopic and macroscopic stages in reactor core analysis. A simple approach implicitly requires too much computation time, and is infeasibile for actual near-future designs. Therefore we developed a new calculation technique that can shorten computation time with minimum degradation in accuracy by changing the solution method of the domination equation according to importance of the domain. In the next section, a typical solution method for the neutron transport equation, the domination equation in the reactor core analysis, will be described.

2.2   Solution Methods of Neutron Transport Equation

The behavior of neutrons is described with the Boltzmann equation, but this is difficult to solve without discretization of energy. Normally, it is treated as a set of simultaneous equations of many groups after integrating within appropriate energy ranges. Our interest is to solve the discretized transport equations and obtain the space and energy distribution of neutron and fission under various conditions.

Several methods can be utilized to solve the discretized transport equations[2, 3]. For example, space phase can be divided as the flight direction of a neutron in the ordinate angle method (S_N method). The Legendre expansion method (P_L method) expresses the angle distribution with polynomial functions. These methods show good accuracy, but their computation costs are quite high. On the other hand, the diffusion approximation method treats only a representative direction, which is a good approximation in general and requires less computation cost. However, accuracy worsens when there is a large gradient on neutron distribution, because anisotropy in the neutron flight direction is not accounted for. As mentioned, neutron currents are discretized in angles for the S_N method and in moments for the P_L method. Under those circumstances, how can we be efficient when calculations meshed with different theories for solution methods reside in adjacent positions? In such a situation, one solution is based on an object-oriented approach.

3   Object-Oriented Approach for Sequential Method

In an object-oriented approach, a system is built using "objects" as if they were bricks that unifly data and procedures. These objects have high independence from each other, so direct reference to internal data of other objects is not allowed in this approach. In other words, one must call procedures in an object to set or retrieve data from the object. This apparently troublesome limitation produces the security of the model, which leads to high modularity, extensibility and reusability.

First, we assume an implementation of the object-oriented approach to a meshing method such as the finite difference method (FDM). In the iterative FDM, successive calculations are performed until all the governing equations are satisfied. Variables in a calculation mesh are determined to satisfy the governing equation locally.

Second, we consider each calculation mesh as an "object." Each calculation node object has parameters necessary for calculation within the node. The

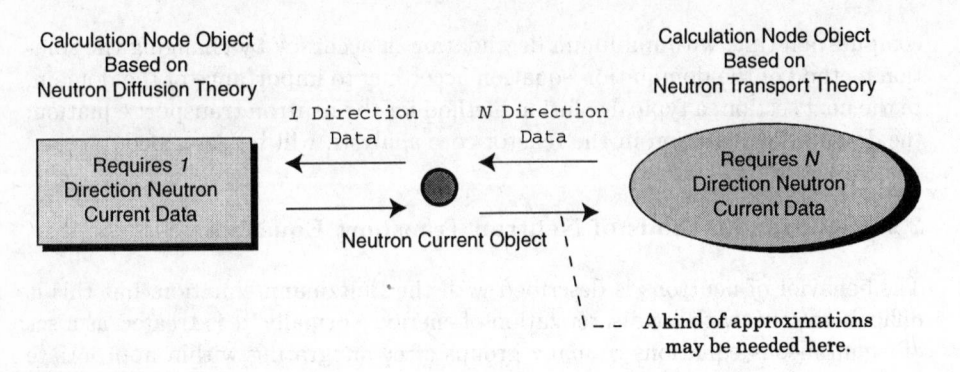

Fig. 1. A neutron current object that has automatic type-conversion inside can connect different type of calculation objects that represent calculation meshes

governing equations are solved locally in the object by a solution method procedure of its own. Information to be exchanged between calculation objects is transferred by an object abstractly, for instance the neutron-current object in our application. Calculation objects can retrieve information from the neutron-current objects with a specified data type by the mechanism of automatic type-conversion that is built in the neutron current object. Therefore, a calculation object does not need to know what kinds of calculation objects exist in adjacent positions. In this manner, various kinds of calculation method can be assigned at arbitrary mesh points in the domain. Figure 1 illustrates this object-oriented approach.

Each calculation node class has a specific procedure based on its calculation method, and the interface to the procedure is the same for all classes, for instance "calc()." Therefore a new class based on another calculation method can be derived from the base calculation class with common properties and procedures by defining the method-specific properties and overloading the "calc()" member function. Some classes of calculation node are listed in Table 1.

Those calculation nodes are stored and managed in the same manner that a container object can be built by the *ContainerController* object with the *Region* object. The container object represents the whole or decomposed domain of the system, while each calculation object in the container object represents a calculation mesh. Note that calculation node objects in the container are independent each other and can be dynamically replaced by other types of calculation nodes (for example, with a higher-order approximation) at any stage during the iterative calculations. In this way, a quite flexible system can be built easily with the object-oriented approach. For example, a calculation method/algorithm can be upgraded by region and by time. The object classes that are used for construction of the calculation domain are listed in Table 2.

Table 1. List of currently available classes of calculation objects

Class Name	Description
Node	Abstract superclass for all kinds of nodes
BNode	Node on system and processor boundary derived from Node class. Basic calculation node with method independent properties and procedures
CNode	for each specific calculation nodes as a superclass. This class is derived from the *Node* class.
FDDNode	Calculation node based on the finite difference diffusion approximation (derived from *CNode*)
SP3Node	Calculation node based on the Simplified P3 transport approximation (derived from *CNode*)
...and more	There are several derived classes of calculation node for each specific calculation method/algorithm.

Table 2. Object classes for constructing calculation domain

Class Name	Description
Region	Abstracted representation of geometric configuration of calculation domain, such as boundary conditions, material map, calculation node map, etc.
Container	Abstracted calculation domain that keeps nodes inside, performs three-dimensional Red/Black checkerboard sweep, etc.
ContainerController	Controlling object for *Container*.

4 Natural Extension to a Parallel Algorithm

With a finer model of computation, the growth in demand for computing resources becomes a significant problem. As a result, attention has been paid to parallel and distributed computing. In this section a natural extension of the above object-oriented approach to a parallel and distributed computing environment is described.

A merit of the object-oriented approach is independence among objects by encapsulating information in them. This allows concurrency among objects because object interaction is minimized by the nature of the object-oriented approach.

Parallel computing on a distributed-memory system requires inter-processor communication over a network. Message passing libraries such as MPI[4, 5] and PVM[6] are provided for high performance computing that supports transmission of arrays of basic data types (int, double, etc) and derived data types such as vectors and structures. However a higher data abstraction by object archiving and transmission is not supported. Therefore we introduced a mechanism to encode and decode objects to be transferred. With this extension, objects are virtually transmitted between processors with great security of data inside objects.

6

In our approach, assignment of portions of a problem to each processor is performed by domain decomposition[7]. Each processor has an object defined by the *Region* class in which all the information is encapsulated to map an actual problem domain into calculation objects. In parallel computing, for instance, each processor in a group has data for only a part of the total system and computes in it as its responsible domain. In the *Region* object, boundary conditions and some information for parallel computing (such as processor index for each direction) are also defined.

In domain decomposition, a quasi-master processor divides a total system into portions of the domain, produces *Region* objects for them, and distributes them to client processors. Each processor receives a *Region* object and constructs a responsible domain. Note that parallelization of serial code can be done quite naturally with relatively small changes, because the fundamental data structures do not change owing to data encapsulation by objects. Additional work was needed for only a few classes: *Region* and *RegionManager* classes, and *ObjectPass* for domain decomposition and object transmission, respectively. A new class, *CurrentVector*, for packing and unpacking of *Current* objects was also introduced for efficient data transmission between processors.

Figure 2 shows domain decomposition using the *Region* object for parallel computing with two processors. The only difference between parallel and serial computing is whether there are pseudo-boundaries between processors where communication is needed. Consequently, the parallel version follows exactly the same process of convergence as the serial version in a iterative calculation.

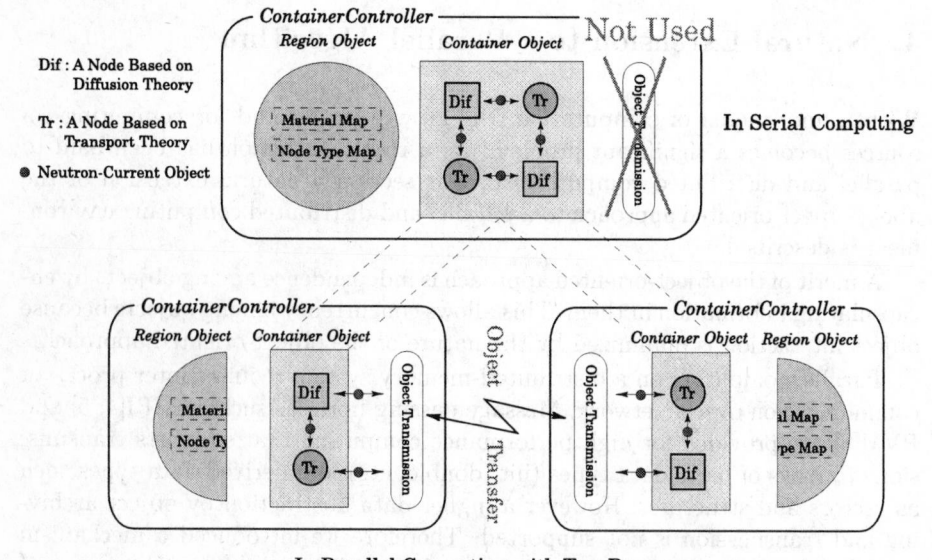

Fig. 2. Parallel computing with two processors by domain decomposition

5 Performance Analysis

A parallel solution algorithm for the neutron transport equation based on this approach has been implemented in C++ as the SCOPE code, currently under developmen[8]. In the SCOPE code, several kinds of calculation nodes with different bases of calculation theory can be used and assigned to arbitrary meshes in the domain. All the classes of calculation nodes currently available in the SCOPE code are listed in Table 3. MPICH[9] was used as the message passing library for communication among processors.

The parallel performance of the SCOPE code was measured on a networked cluster of PCs connected with a Ethernet switch. The cluster consists of a server node and client nodes as described in Table 4.

The server node provides services such as NFS so that the client nodes can mount home directories. This parallel environment can be considered as homogeneous because the actual computation is performed only on the client nodes. The effective performance of peer-to-peer communication between client nodes was reasonable: Measurement of file transfer rates by rcp gave a network bandwidth of about 900KB/s.

Table 3. List of currently available classes of calculation objects

Class Name	Calculation Theory	Number of data in a neutron current object	Relative Computation Load
FDDNode	Diffusion Theory	1	1
SP2Node	Simplified P2	1	1.2
SP3Node	Simplified P3	2	2
SP5Node	Simplified P5	3	3
ANMNode	Analytical Nodal Polynomial Expansion	1	10
S4Node	Discrete Ordinate S4	12	12
S6Node	Discrete Ordinate S6	24	24
S8Node	Discrete Ordinate S8	40	40

Table 4. System configuration of the networked PC cluster

	Server Node	Client Node
Model	Dell OptiPlex Gxa 333	Compaq DeskPro 590
Operating System	Linux 2.0.33	Linux 2.0.33
CPU	Pentium II 333MHz	Pentium 90MHz
RAM	160MB	32–128MB
Network Interface Card	3Com 950	PCNet32

Table 5. Problems analyzed on networked PC cluster

Case Name	Mesh Size	Node Type
Med	60 x 60 x 6	*FDDNode*
Big	60 x 60 x 18	*FDDNode*
Med-sp3	60 x 60 x 6	*SP3Node*
Big-sp3	60 x 60 x 18	*SP3Node*

Performance measurement was done with four problems listed in Table 5, changing the problem size and calculation node type. Speedup curves are shown in Fig. 3 for each case.

In using *FDDNode* in a small problem(Fig. 3-(a)), performance rapidly degrades with the number of processors because of the finer granularity. Problem "Big" is three times larger than "Med" in the z-axis, which improves performance (Fig. 3-(b)). *SP3Node*, with a computational load about twice that of *FDDNode*, showed better performance even in the "Med" case (Fig. 3-(c)). However the performance on eight processors worsened because of the 3D domain decomposition that requires additional initialization and smaller packets for communication as compared to 2D or 1D decompositions. In this case, the 3D domain did not have enough computation load compared to communication load[10]. The last case, using *SP3Node* on "Big" showed good performance on eight processors (Fig. 3-(d)).

Figure 4 predicts efficiency as a function of the number of processors when 2D or 3D decomposition is performed. The curve in the figure gives efficiency relative to perfect speedup. If 50% is set as the criterion for parallel efficiency, one can roughly estimate that about 18 processors can be used for the problem. So good performance can be expected by using 18 or fewer processors in the calculation.

6   Discussion

Good performance and shorter computation time can be expected with parallel computing on a networked PC cluster when the problem to be solved has large enough granularity. The other approach, in contrast, is to use several kinds of calculation node objects in the same computation. For instance, one may assign more accurate calculation node objects for important areas and approximated calculation node objects for less important areas in the domain. This approach can reduce computation time greatly and main good accuracy overall.

As an experiment for this approach, we performed some benchmarks[11]. Three kinds of node configurations were examined: all *FDDNodes*, all *SP3Nodes*, and a hybrid use of *SP3Nodes* and *FDDNodes*. In the last case, *SP3Nodes* were assigned to important regions such as fuel and control rods), while *FDDNodes* were used for less important region such as peripheral moderators. Calculation results and relative computation times are listed in Table 6. The last case, the

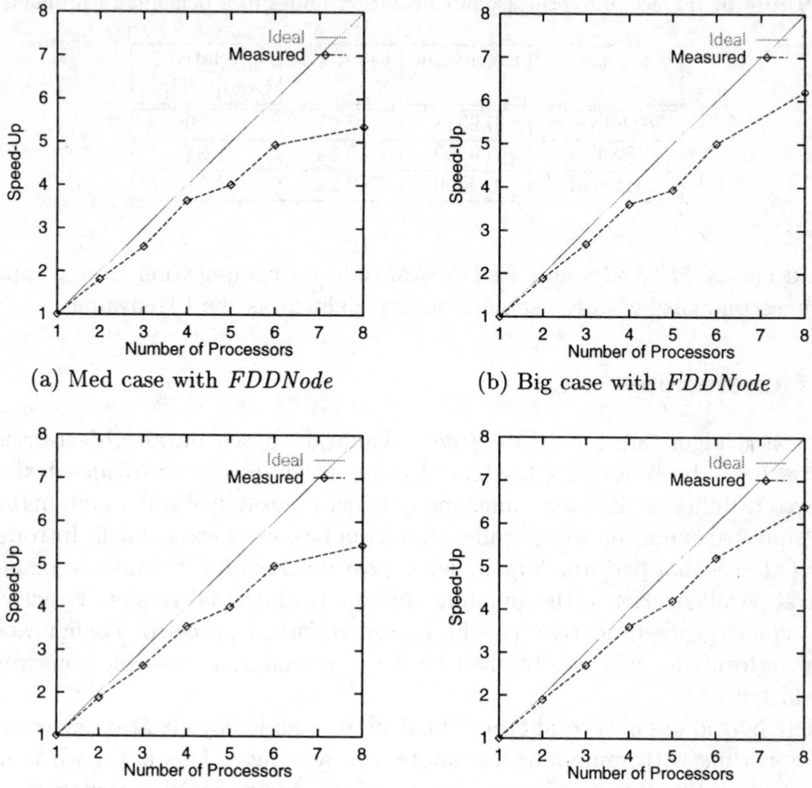

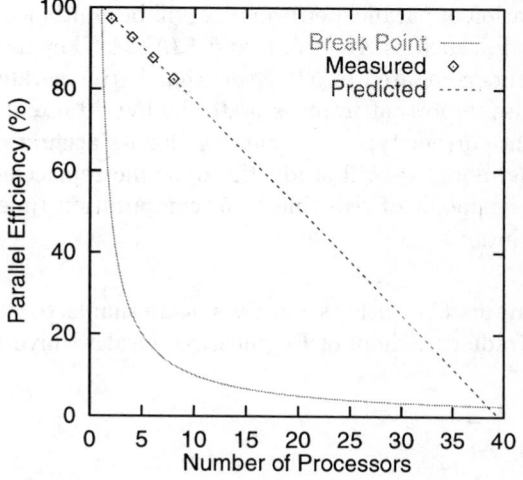

(a) Med case with *FDDNode*

(b) Big case with *FDDNode*

(c) Med case with *SP3Node*

(d) Big case with *SP3Node*

Fig. 3. Speedup curves for the SCOPE code on the networked PC cluster

Fig. 4. Prediction of parallel efficiency in case "Big" with *SP3Node*

Table 6. Impact of hybrid use of calculation nodes in a benchmark problem

Case	Eigenvalue	Error (%)	Relative Comp. Time
FDDNode	0.9318	-2.97	1.00
SP3Node	0.9590	-0.14	1.38
Hybrid	0.9590	-0.14	1.11

hybrid use of *SP3Node*s and *FDDNode*s, reduced computation time compared to the second case with the same accuracy, defined as the Eigenvalue.

7 Conclusions

A solution algorithm for an iterative calculation based on an object-oriented approach has been developed. The object-oriented approach provides flexibility and extensibility at the same time, and enhances reusability and maintainability.

A natural extension to a parallel algorithm has also been studied. Introduced object classes that perform domain decomposition and object transmission help a natural parallelization with minimum changes to the serial version. Exactly the same convergence properties can be expected for all processor configurations. High performance can be obtained for large granularity, even on a networked PC cluster.

The hybrid use of several types of calculation node objects also reduces computation time with minimum degradation of accuracy. This approach is quite attractive and further studies are expected introducing multigrid analysis.

8 Future Study

Further investigation of parallel performance will be continued using other kinds of calculation nodes, such as *ANMNode* and *S4Node*. They have heavier computation requirements compared to *SP3Node*, thus better parallel speedup can be expected. It is also important to investigate the hybrid use of calculation nodes requiring different current types, P_L and S_N, for its accuracy and parallel performance. Furthermore, we will study the dynamic replacement of calculation nodes from the viewpoint of reducing total computation time and maintaining the quality of solutions.

Acknowledgement The authors express sincere thanks to Professor Toshikazu Takeda of the Graduate School of Engineering, Osaka University for invaluable discussion.

References

[1] Stamm'ler R. J. J. and Abbate M. J. *Methods of Steady-State Reactor Physics in Nuclear Design*. Academic Press, 1983.

[2] Lewis E. E. and Miller W. F. Jr. *Computational Methods of Neutron Transport.* Wiley-Interscience, 1983.

[3] Ronen Y. *Handbook of Nuclear Reactors Calculation Vol. I.* CRC Press, 1986.

[4] Gropp W., Lusk E., and Skjellum A. *Using MPI: Portable Parallel Programming with the Message Passing Interface.* MIT Press, 1994.

[5] Snir M., Otto S., Lederman S. H., Walker D., and Dongarra J. *MPI: The Complete Reference.* MIT Press, 1996.

[6] Geist A., Beguelin A., Dongarra J., Jiang W., Manchek R., and Sunderam V. *PVM: Parallel Virtual Machine A Users' Guide and Tutorial for Networked Parallel Computing.* MIT Press, 1994.

[7] Foster I. *Designing and Building Parallel Programs.* Addison Wesley, 1995.

[8] Tatsumi M. and Yamamoto A. Scope: A scalable and flexibile parallel algorithm based on object-oriented approach for core calculations. *Joint Int. Conf. on Mathematical Methods and Supercomputing for Nuclear Applications*, October 6-10, Saratoga Springs, New York 1997.

[9] MPICH Web Home Page available at http://www.mcs.anl.gov/mpi/mpich/.

[10] Hanebutte U. R. and Tatsumi M. Study of parallel efficiency in message passing environments. *High Performance Computing '96*, April 8-11, New Orleans, Louisiana 1996.

[11] Takeda T. and Ikeda H. 3-d neutron transport benchmarks. *NEACRP-L-330*, 1991.

An Object-Oriented Time Warp Simulation Kernel*

Radharamanan Radhakrishnan, Dale E. Martin, Malolan Chetlur, Dhananjai
Madhava Rao, and Philip A. Wilsey

University of Cincinnati, Cincinnati, OH, USA

Abstract. The design of a Time Warp simulation kernel is made difficult by the inherent complexity of the paradigm. Hence it becomes critical that the design of such complex simulation kernels follow established design principles such as object-oriented design so that the implementation is simple to modify and extend. In this paper, we present a compendium of our efforts in the design and development of an object-oriented Time Warp simulation kernel, called WARPED. WARPED is a publically available Time Warp simulation kernel for experimentation and application development. The kernel defines a standard interface to the application developer and is designed to provide a highly configurable environment for the integration of Time Warp optimizations. It is written in C++, uses the MPI message passing standard for communication, and executes on a variety of platforms including a network of SUN workstations, a SUN SMP workstation, the IBM SP1/SP2 multiprocessors, the Cray T3E, the Intel Paragon, and IBM-compatible PCs running Linux.

1 Introduction

The Time Warp parallel synchronization protocol has been the topic of research for a number of years, and many modifications/optimizations have been proposed and analyzed [1, 2]. However, these investigations are generally conducted in distinct environments with each optimization re-implemented for comparative analysis. Besides the obvious waste of manpower to re-implement Time Warp and its affiliated optimizations, the possibility for a varying quality of the implemented optimizations exists.

The WARPED project is an attempt to make a freely available object-oriented Time Warp simulation kernel that is easily ported, simple to modify and extend, and readily attached to new applications. The primary goal of this project is to release an object-oriented software system that is freely available to the research community for analysis of the Time Warp design space. In order to make WARPED useful, the system must be easy to obtain, available with running applications, operational on several processing platforms, and easy to install, port, and extend.

* Support for this work was provided in part by the Advanced Research Projects Agency under contracts DABT63–96–C–0055 and J–FBI–93–116.

This paper describes the general structure of the WARPED kernel and presents a compendium of the object-oriented design issues and problems that were required to be solved. In addition, a description of two distinct application domains for WARPED is presented. WARPED is implemented as a set of libraries from which the user builds simulation objects. The WARPED kernel uses the MPI [3] portable message passing interface and has been ported to several architectures, including: the IBM SP1/SP2, the Cray T3E, the Intel Paragon, a network of SUN workstations, an SMP SUN workstation, and a network of Pentium Pro PCs running Linux.

The WARPED system is implemented in C++ and utilizes the object-oriented capabilities of the language. Even if one is interested in WARPED only at the system interface level, they must understand concepts such as inheritance, virtual functions, and overloading. The benefit of this type of design is that the end user can redefine and reconfigure functions without directly changing kernel code. Any system function can be overloaded to fit the user's needs and any basic system structure can be redefined. This capability allows the user to easily modify the system queues, algorithms or any part of the simulation kernel. This flexibility makes the WARPED system a powerful tool for Time Warp experimentation.

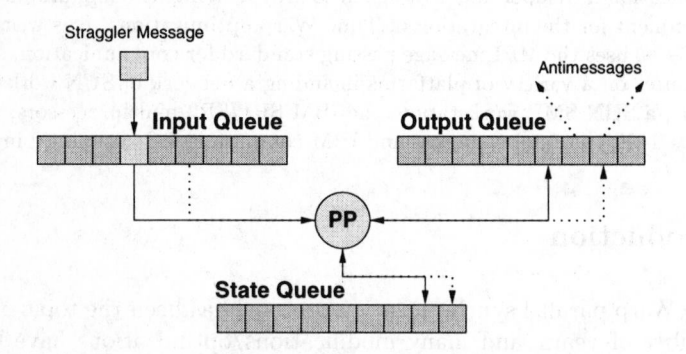

Fig. 1. A logical process in a Time Warp simulation

Another benefit of the object-oriented nature of the WARPED application interface is that by its very design it is simple to "plug in" a different kernel. A sequential simulation kernel is supplied in the WARPED distribution in addition to the Time Warp kernel. Version 0.9 of the WARPED is available via the www at http://www.ece.uc.edu/~paw/warped/. The remainder of this paper is organized as follows. Section 2 presents a description of the Time Warp paradigm. Section 3 details the WARPED kernel's application/kernel interface and presents a compendium of the design issues that were required to be solved for the development of the WARPED system. Section 4 demonstrates, through two examples, the construction of simulation applications using the WARPED kernel. Finally, Sect. 5 contains some concluding remarks.

2 Background

In a Time Warp synchronized discrete event simulation, *Virtual Time* [2] is used to model the passage of the time in the simulation. The virtual time defines a total order on the events of the system. The simulation state (and time) advances in discrete steps as each event is processed. The simulation is executed via several simulator processes, called simulation objects or logical processes (LP). Each LP is constructed from a physical process (PP) and three history queues. Figure 1 illustrates the structure of an LP. The input and the output queues store incoming and outgoing events respectively. The state queue stores the state history of the LP. Each LP maintains a clock that records its Local Virtual Time (LVT). LPs interact with each other by exchanging time-stamped event messages. Changes in the state of the simulation occur as events are processed at specific virtual times. In turn, events may schedule other events at future virtual times.

The LPs must be synchronized in order to maintain the causality of the simulation; although each LP processes local events in their (locally) correct time-stamp order, events are not globally ordered. Fortunately, each event need only be ordered with respect to events that affect it (and, conversely, events that it affects); hence, only a partial order of the events is necessary for correct execution [4]. Under optimistically synchronized protocols (*e.g.,* the Time Warp model [2]), LPs execute their local simulation autonomously, without explicit synchronization. A causality error arises if a LP receives a message with a time-stamp earlier than its LVT (a *straggler* message). In order to allow recovery, the state of the LP and the output events generated are saved in history queues as events are processed. When a straggler message is detected, the erroneous computation must be undone—a *rollback* occurs. The rollback process consists of the following steps: the state of the LP is restored to a state prior to the straggler message's time-stamp, and then erroneously sent output messages are canceled (by sending *anti-messages* to nullify the original messages). The global progress time of the simulation, called Global Virtual Time (GVT), is defined as the time of the earliest unprocessed message in the system [1, 5, 6]. Periodic GVT calculation is performed to reclaim memory space as history items with a time-stamp lower than GVT are no longer needed, and can be deleted to make room for new history items.

3 The WARPED Application and Kernel Interface

The WARPED kernel presents an interface to the application from building logical processes based on Jefferson's original definition [2] of Time Warp. Logical processes (LPs) are modeled as entities which send and receive events to and from each other, and act on these events by applying them to their internal state. This being the case, basic functions that the kernel provides to the application are methods for sending and receiving events between LPs, and the ability to specify different types of LPs with unique definitions of state. One departure

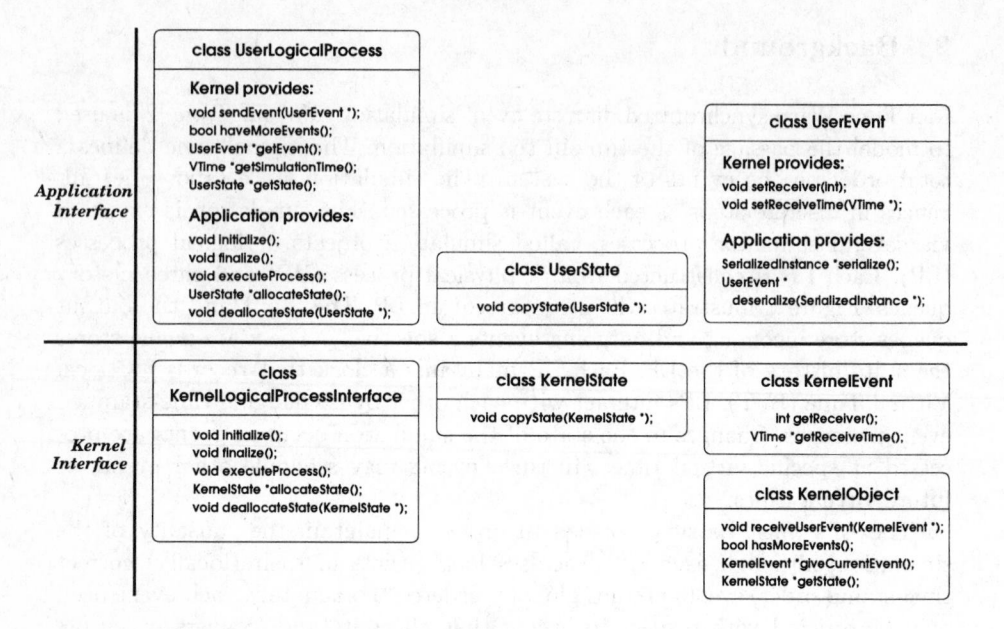

Fig. 2. Application and kernel interface

from Jefferson's presentation of Time Warp is that LPs are placed into groups called "clusters". LPs on the same cluster communicate with each other without the intervention of the message system, which is *much* faster than communication through the message system. Hence, LPs which communicate frequently should be placed on the same cluster. Another feature of the cluster is that it is responsible for scheduling the LPs. Note that the LPs within a cluster operate as Time Warp processes; even though they are grouped together, they aren't coerced into synchronizing with each other.

Control is passed between the application and the kernel through the cooperative use of function calls. This means that when a function is called in application code, the application is not allowed to block for any reason. Since the application has control of the single thread of control through its cluster, it could end up waiting forever. In order for the kernel to correctly interact with the application code, the user must provide several functions to the kernel. These functions define such things as how to initialize each LP, and what each LP does during a simulation cycle. In addition, if the user would like to use a nonstandard definition of time, facilities are in place to provide a user-defined time class to the kernel. By default, WARPED has a simple notion of time. More precisely, time is defined in the class VTime as a signed integer. Obviously, particular instances of an application may have different requirements for the concept of time. For example, simulators for the hardware description language VHDL [7] require a more complex definition of time. If the simple, kernel-supplied version of time is not sufficient, the application programmer must define the class VTime with data members appropriate to the application's needs. In addition, the user must use the preprocessor macro USE_USER_VTIME during compilation. The WARPED kernel also has requirements about the defined methods of the type VTime. Specifically, the implementation of VTime must supply the following operators and data, either by default or through explicit instantiation:

- Assignment (=), Addition (+), and subtraction (−) operators.
- The relational operators: ==, !=, >=, <=, >, <.
- Constant objects ZERO, PINFINITY, and INVALID_VTIME of type VTime, which define, respectively, the smallest, largest, and invalid time values.
- INVALID_VTIME must be less than ZERO.
- The insertion operator (<<) for class ostream, for type VTime.

The application interface is implemented through the object-oriented features of the C++ language. The simulation kernel is built from several classes, allowing the user to define a system configuration by specifying the classes to use, without rewriting system code. Application specific code is derived from the WARPED kernel. This allows application code to transparently access kernel functions and is restrictive enough to hide communication and Time Warp details from the user. This section describes what is necessary for an application writer to provide the WARPED kernel, and what the simulation kernel provides to the application in return. To use the WARPED kernel, the application programmer must provide three class definitions corresponding to the logical process (LP), the notion of state for that LP, and a definition (or definitions) for events.

LPs form the core of the discrete event simulation. An LP represents an entity that can send/receive events to/from other LPs. As a result of these events, changes are made to the LP's internal state (and output may result). Figure 2 illustrates the application and the kernel interfaces presented by the WARPED system. The interface as seen by an user's LP is represented by the UserLogicalProcess class definition. The class definition is divided into two parts. The first part is the set of methods that the kernel provides to the LP. These methods are provided by the kernel to the LP for communication (sendEvent, getEvent), querying the kernel for information (haveMoreEvents, getSimulationTime) and for accessing its state (getState). In addition to these methods, there are some internal methods that the kernel calls periodically. These include message polling primitives to check for the arrival of messages from remote processors and garbage collection primitives. The second part consists of a set of methods that the application writer overrides. The kernel will call these methods at various times through out the simulation. Each method in this set has a specific function. The initialize method gets called on each LP before the simulation begins. This gives each LP a chance to perform any actions required for initialization. For example, initialization might include opening files, setting up the initial state of an LP or the transmission of initial setup events to the distributed processes in the simulation. Conversely, the method finalize is called after the simulation has ended. This allows the LPs to "clean up" after themselves, perform actions such as closing files, compute statistics, and produce output. The method executeProcess of an LP is called by the kernel whenever the LP has at least one event to process. The kernel calls allocateState in an

LP when it needs the LP to allocate a state on its behalf. `deallocateState` is called by the kernel to hand back a state to the application when it is done with it. At this point, the application may deallocate it, or store it for later use.

Any LP will have some state that needs to be defined. The LP modifies its state in response to various events that it receives. This behavior is completely user application specific and the application must define certain methods related to state for the simulation kernel to call. These methods include the creation and the duplication of the state. Figure 2 illustrates the user application's interface to the state. The method `copyState` is called by the kernel to copy the data from the `UserState` into a newly created state which is then archived (for rollback recovery purposes). This method must be overridden by the user application. If the application's definition of state contains no pointers, then a bitwise copy is adequate for this method. If the application contains pointers in its state, or objects that contain pointers, then this method has to take appropriate actions to copy the pointers "correctly", as defined by the needs of the application. This is necessary because the kernel has no knowledge about the user application's state.

Events represent the communication between the LPs. Figure 2 illustrates the definition of the `UserEvent` class. Once again, the definition is a two part definition wherein one set of methods is provided by the kernel and the other set is overridden by the application writer. The method `setReceiver` allows the application to set the simulation id of the receiving LP [1]. `setReceiverTime` allows the application to set the simulation time that this event should be received at. The methods `serialize` and `deserialize` are provided so that the application may maintain architectural transparency and portability among events. It is also necessary for checkpointing in optimistic fossil collection [8] and failure recovery.

The design of the WARPED API was motivated by several design issues. These issues were central to the object-oriented design of the system and needed to be solved for constructing a simple and extensible programming interface. For example, when the kernel needs information about data structures within the application, they can be passed into the kernel in two ways : through template classes or through virtual interface methods. One example of this is the state class definition. The user state can be passed into the kernel through templates. All that is required is that the `UserLogicalProcess` class be templatized on `UserState`. However, to reduce overall compilation time, static executable size and facilitate the use of different types of states, the templatization approach was avoided. The convention currently followed is to have the LP and the kernel share the responsibility of allocating, maintaining and deallocating the state through the use of virtual methods. Although the common perception in scientific computing is that abstraction is the enemy of performance, we have found that the extensive use of virtual methods and other abstractions does not drastically affect performance. When kernel data or functions need to be made available to the user, they can be accessed by one of two mechanisms: through the C++

---

[1] As it is the user's responsibility to register an LP with a unique simulation id, the application can use the `setReceiver` method to connect LPs together.

inheritance mechanism (classes that the user defines must be derived from kernel defined classes), and through "normal" function calls to methods defined by objects in the WARPED kernel.

In addition, avoiding the use of templates facilitates the distribution of source code as stand alone libraries which do not require recompilation. This enables the development of "object factories" by independent vendors. With object factories, vendors can permit different users to use various components from their object factories without revealing the source code. To enable this type of "plug-and-play", C++ *composition* was carried out in preference to *inheritance* in the source code. Composition also helps in achieving dynamic algorithm/method reconfiguration (*i.e.*, reconfiguration "on-the-fly" without recompilation).

Also, avoiding the use of templates makes the WARPED system simpler to port to different compilers on different architectures. To achieve interoperability on heterogeneous platforms, the serialization and deserialization operations play a vital role. Currently serialization and deserialization of events as well as states is supported. These operations are invoked only when events or states cross architecture boundaries. Serialization and deserialization is also applied to checkpointing to facilitate failure recovery.

In the current version of WARPED, there are several Time Warp implementation optimizations that can be turned on/off. A configuration file is used to allow the user to change between the options of the simulation kernel at compile time. These options fall under several broad categories: Schedulers, Fossil Managers, State Managers, Memory Managers, and Time Warp optimizations (such as dynamic cancellation [9], dynamic checkpointing [10] and dynamic message aggregation [11]). The user specifies a selection from this set of options and compiles this selection. A better way to implement this is through dynamic configuration. Each optimization is implemented as a specific function and at run-time, a simulation object (or some central configuration object) can dynamically select and reconfigure (through function pointers) the optimization and switch between optimizations if the need arose [12].

4 Applications for WARPED

Several applications have already been developed that use the WARPED kernel. These applications primarily belong to two application domains: a queuing model simulation library called KUE, and TYVIS, a simulation kernel for the VHDL hardware description language [7]. KUE is a simple package developed for debugging, testing, and initial profiling of WARPED and any extensions thereof. TYVIS is a larger package designed to stress the simulation kernel with large examples of digital systems. It also demonstrates the extensibility of the WARPED kernel. The developers hope that other investigators will implement additional applications with WARPED which they can include as part of the distribution. As space constraints prevent us from presenting the performance of the WARPED system, the rest of this section is devoted to the description of WARPED applications.

4.1 KUE: A Queuing Model Library

The KUE system is a library of queuing models built on top of the WARPED kernel. KUE is a set of C++ classes that enable the creation of parallel queuing applications. XKUE is a TCL/TK front end for queue to allow "point and click" creation of queuing models. The KUE library contains class definitions of seven different queuing objects (source, fork, join, delay, queue, server and sink objects). Each object class definition encapsulates the functionality of the queuing object in accordance with the WARPED interface. Two examples are distributed with the WARPED kernel that make use of the KUE libraries.

The first, SMMP, is designed to simulate several processors, each with their own cache, and sharing a global memory. The model is generated by a program which lets the user adjust the following parameters: the number of processors/caches to simulate, the number of LPs to generate, the speed of cache, the speed of main memory, and the cache hit ratio. The second example, RAID, is a simulation of a nine disk RAID level 5 array of IBM 0661 3.5" 320MB SCSI drives with a flat-left symmetric parity placement policy. Sixteen processes generate requests for data stripes of random lengths and locations. These requests are sent to fork processes which split them into specific disk-level requests according to the RAID placement policy. The nine server processes, one per simulated disk, process the requests in a first-come first-served fashion. After processing each request, the disks route their responses back to the originating processes. Both these sample queuing applications posses class definitions that derive from the seven basic queuing model definitions in the KUE library. Further details regarding these applications are available in the literature [12].

4.2 TyVIS: A Parallel VHDL Simulation Kernel

The TyVIS VHDL simulation kernel was designed to take advantage of the object-oriented design of the WARPED kernel. It requires no modifications to the kernel, yet extends WARPED with full VHDL simulation capability (as described in [7]). Its implementation takes advantage of several design features of WARPED, and even reuses some of WARPED's basic classes for TyVIS's internal data structures. The main class of TyVIS is `VHDLKernel`, which is derived from the `UserLogicalProcess` class.

The semantics of VHDL require that certain events generated during a simulation cycle not be applied to a signal's value, based upon each event's timestamp. This process is called marking, and is best implemented with a time-ordered queue. Rather than writing an entirely new data structure, the `OutputQueue` class of the WARPED distribution was reused, becoming a base class for the `MarkedQueue` class. The public interface to `MarkedQueue` is identical to that of `OutputQueue`; all additional data members and methods are private. This reuse of the existing code allowed the `MarkedQueue` class to be written and debugged in a matter of a few hours. Also, since `MarkedQueue` only accesses the public interface of `OutputQueue`, any changes in the implementation of `OutputQueue` will be transparent.

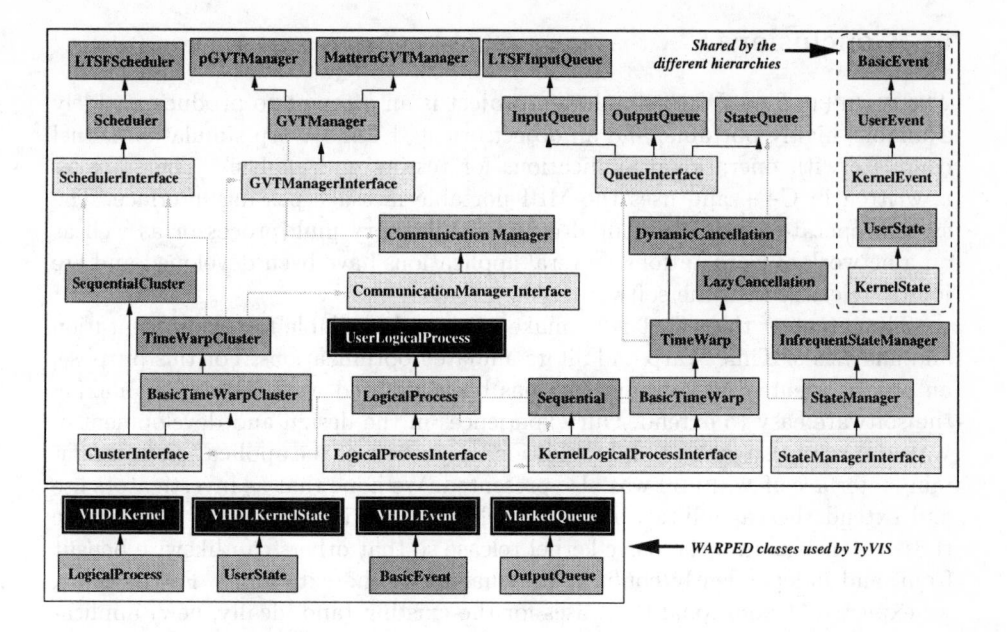

Fig. 3. A synopsis of WARPED's class derivation hierarchy

Each VHDL process has a unique state class which defines the VHDL signals and local variables that the process can access. This state class is built from WARPED's `UserState` class with the necessary user-defined methods. This allows the Time Warp functions of state queuing, rollbacks, and garbage collection to proceed normally. The only requirement to the state class for this is that the class define `operator=`.

Processes are invoked by calling `VHDLKernel::executeProcess()`, which overrides the similar method in the `UserLogicalProcess` class. This method updates LVT and applies all events in the input queue occurring on any signals contained in the process at the current time. The specific VHDL process code is then executed by calling the object's `executeVHDL` method, supplied by the user. When the process returns control to the VHDL kernel, the kernel then determines which newly generated events need to be transmitted to other processes, and transmits them, using the `sendEvent` call from the WARPED kernel. Eventually, control is returned to WARPED. If a process is rolled back, the VHDL kernel never knows about it, since all related processing is contained entirely in the WARPED code, lower down in the derivation hierarchy. Complete replacement of the WARPED kernel with a conservatively synchronized simulation kernel would have no effect on TYVIS; it is completely isolated from whatever processing is performed by WARPED. A synopsis of the WARPED class derivation tree is illustrated in Fig. 3. Base class definitions form the root of the derivation hierarchy. Figure 3 also depicts the WARPED classes reused by the TYVIS VHDL simulation kernel.

5 Conclusions

The WARPED Time Warp simulation project is an attempt to produce a widely available, highly portable, and an object-oriented Time Warp simulation kernel complete with operational applications for testing and analysis. The software is written in C++ and uses the MPI portable message passing interface. The system operates on a distributed or shared memory multiprocessor as well as on a network of workstations. Several applications have been developed and are jointly released with the software.

The intent of this effort is to make a testbed available for experimentation and analysis of Time Warp and all its affiliated optimizations. For this purpose, an object-oriented design approach has been followed with the aim of making the software easy to extend. Our experiences in the design and development of WARPED were also presented. In addition, a synopsis of the application programming interface of WARPED was also presented. We hope that as investigators use and extend the capabilities of the kernel that we will be allowed to integrate those extensions into the basic kernel release so that others can likewise benefit from, and independently confirm, the analysis of the extensions. Furthermore, we expect that additional test cases for the existing (and ideally, new) applications will be independently developed and submitted for inclusion into the kernel release (and thereby promoting reuse of source code).

References

[1] R. Fujimoto. Parallel discrete event simulation. *Communications of the ACM*, 33(10):30–53, October 1990.

[2] D. Jefferson. Virtual time. *ACM Transactions on Programming Languages and Systems*, 7(3):405–425, July 1985.

[3] W. Gropp, E. Lusk, and A. Skjellum. *Using MPI: Portable Parallel Programming with the Message-Passing Interface*. MIT Press, Cambridge, MA, 1994.

[4] L. Lamport. Time, clocks, and the ordering of events in a distributed system. *Communications of ACM*, pages 558–565, July 1978.

[5] Yi-Bing Lin. Memory management algorithms for optimistic parallel simulation. In *6th Workshop on Parallel and Distributed Simulation*, pages 43–52. Society for Computer Simulation, January 1992.

[6] F. Mattern. Efficient algorithms for distributed snapshots and global virtual time approximation. *Journal of Parallel and Distributed Computing*, 18(4):423–434, August 1993.

[7] *IEEE Standard VHDL Language Reference Manual*. New York, NY, 1993.

[8] C. H. Young and P. A. Wilsey. Optimistic fossil collection for Time Warp simulation. In H. El-Rewini and B. D. Shriver, editors, *29th Hawaii International Conference on System Sciences (HICSS-29)*, volume Volume I, pages 364–372, January 1996.

[9] R. Rajan, R. Radhakrishnan, and P. A. Wilsey. Dynamic cancellation: Selecting Time Warp cancellation strategies at runtime. *VLSI Design*, 1998. (forthcoming).

[10] J. Fleischmann and P. A. Wilsey. Comparative analysis of periodic state saving techniques in Time Warp simulators. In *Proc. of the 9th Workshop on Parallel and Distributed Simulation (PADS 95)*, pages 50–58, June 1995.

[11] M. Chetlur, N. Abu-Ghazaleh, R. Radhakrishnan, and P. A. Wilsey. Optimizing communication in Time-Warp simulators. In *12th Workshop on Parallel and Distributed Simulation*. Society for Computer Simulation, May 1998.

[12] R. Radhakrishnan, N. Abu-Ghazaleh, M. Chetlur, and P. A. Wils ey. On-line configuration of a Time Warp parallel discrete event si mulator. In *1998 International Conference on Parallel Processing, (ICPP'98)*. IEEE Computer Society Press, August 1998. (forthcoming).

Particle Beam Dynamics Simulations Using the POOMA Framework[*]

William Humphrey, Robert Ryne, Timothy Cleland, Julian Cummings, Salman Habib, Graham Mark, and Ji Qiang

Los Alamos National Laboratory, Los Alamos, NM, USA

Abstract. A program for simulation of the dynamics of high intensity charged particle beams in linear particle accelerators has been developed in C++ using the POOMA Framework, for use on serial and parallel architectures. The code models the trajectories of charged particles through a sequence of different accelerator beamline elements such as drift chambers, quadrupole magnets, or RF cavities. An FFT-based particle-in-cell algorithm is used to solve the Poisson equation that models the Coulomb interactions of the particles. The code employs an object-oriented design with software abstractions for the particle beam, accelerator beamline, and beamline elements, using C++ templates to efficiently support both 2D and 3D capabilities in the same code base. The POOMA Framework, which encapsulates much of the effort required for parallel execution, provides particle and field classes, particle-field interaction capabilities, and parallel FFT algorithms. The performance of this application running serially and in parallel is compared to an existing HPF implementation, with the POOMA version seen to run four times faster than the HPF code.

1 Introduction

Particle accelerators have played a central role in shaping our present understanding of the fundamental nature of matter. At the same time, the application of accelerator theory and technology has contributed to substantial progress in other branches of science and technology. This historical trend is expected to continue with particle accelerators playing an increasingly important role in basic and applied science. As examples of recent applications, many countries are now involved in efforts aimed at developing accelerator-driven technologies for transmutation of radioactive waste, disposal of plutonium, energy production, and production of tritium. Additionally, next-generation spallation neutron sources based on similar technology will play a major role in materials science and biological science research. Finally, other types of accelerators such as the Large Hadron Collider (LHC), the International Linear Collider (ILC), and fourth-generation light sources will have a major impact on basic and applied scientific research.

[*] This work was performed under the auspices of the U.S. Department of Energy by Los Alamos National Laboratory under Contract No. W-7405-Eng-36.

For all of these projects, high-resolution modeling far beyond that which has ever been performed by the accelerator community is required to reduce cost and technological risk, and to improve accelerator efficiency, performance, and reliability. Indeed, such modeling is essential to the success of many of these efforts. For example, high average power linear accelerators, such as those needed for tritium production, must operate with extremely low beam loss (~ 0.1 nA/m) to prevent unacceptably high levels of radioactivity. To ensure that this requirement will be met, it is necessary to perform very high-resolution simulations using on the order of 100 million particles in which the beam propagates through kilometers of complicated accelerating structures. These simulations can only be performed on the most advanced high performance computing platforms using software and algorithms targeted to parallel and distributed environments. The calculations require performance of hundreds of GFLOPS to TFLOPS, and core memory requirements of hundreds of GBytes.

The beam dynamics modeling effort has concentrated so far on parallel calculations for the design of proton linear accelerators (linacs). Such accelerators are the machines of choice for applications including radioactive waste treatment and tritium production. Two-dimensional and fully three-dimensional beam dynamics codes that take into account both external accelerating and focusing fields, as well as the inter-particle Coulomb forces in the beam are in an advanced stage of development and have already been used for accelerator design studies [1, 2]. This paper describes the design and implementation of a parallel application used to model high-intensity charged particle beams moving through a linear accelerator, using an object-oriented design in C++ based on the POOMA Framework [3, 4]. The performance of this code is compared to an HPF implementation of the application, running serially and in parallel on the SGI Origin2000 parallel computers available at Los Alamos National Laboratory.

2   Simulating Linear Accelerators

To simulate the motion of charged particles through a linear accelerator, we have employed an object-oriented (OO) software design in our application. Using an OO design strategy makes it easier to develop modular, maintainable code which can easily be extended to incorporate new algorithms, simulation components, and capabilities. The characteristics of linear accelerators, consisting of sequences of beamline elements through which particles move as they are accelerated, lend themselves quite well to being modeled using an OO design. We can consider this system as being comprised of the following abstractions.

Beamline Elements consist of the distinct portions of the linear accelerator beamline through which the particles move. Particles interact with the elements in various ways as they propagate through them; for example, quadrupole magnet elements focus the beam as the charged particles move through their magnetic fields.

The Beamline comprises the collection of different beamline elements which make up the linear accelerator, in the order the elements are encountered by the particles.

The Beam is the set of charged particles being accelerated by the system. Particles have characteristics such as phase-space coordinates, charge, and mass, and move through the beamline subject to the equations of motion for a linear accelerator.

The Accelerator is the entire system, comprising the beamline and the beam.

As the particles in the beam move through the beamline, passing through each beamline element, they experience both external forces due to the element they are passing through and internal forces due to the space-charge interaction of the particles with each other. The space-charge forces are calculated using a standard FFT-based particle-in-cell (PIC) algorithm for a collisionless system [5, 6]. In this algorithm, we first solve the Poisson equation

$$\nabla^2 \phi(\boldsymbol{r}) = 4\pi\rho(\boldsymbol{r}) \tag{1}$$

to find the electrostatic potential $\phi(\boldsymbol{r})$ from the charge density field $\rho(\boldsymbol{r})$ of the particles. From $\phi(\boldsymbol{r})$, the space-charge force $\boldsymbol{F}_i(\boldsymbol{r})$ on each particle with charge q_i is computed using

$$\boldsymbol{E}(\boldsymbol{r}) = -\nabla\phi(\boldsymbol{r}) \tag{2}$$
$$\boldsymbol{F}_i(\boldsymbol{r}) = q_i\boldsymbol{E}(\boldsymbol{r}). \tag{3}$$

The standard PIC algorithm, used in codes discussed here, may be summarized as:

1. Scatter charge onto a grid to obtain a discretized charge density $\rho(\boldsymbol{r})$;
2. Solve (1) to determine the electrostatic potential $\phi(\boldsymbol{r})$ on a grid;
3. Compute the electric field vectors $\boldsymbol{E}(\boldsymbol{r})$ from (2) on a grid by finite difference methods;
4. Gather the electric field vectors from the grid to the particle positions, and calculate the force on each particle $\boldsymbol{F}_i(\boldsymbol{r})$ using (3).

The beamline element forces and the space-charge interaction forces result in changes to the momentum and position of the particles, causing them to accelerate through the beamline.

3 Implementation Using the POOMA Framework

Figure 1 presents an overview of the object-oriented design of the particle accelerator simulation code, illustrating the abstractions for the accelerator, beam, and beamline components. Each solid box represents an object; the top half of each box indicates the object name, while the bottom half indicates the important methods or variable for the object. Lines terminating in arrows indicate inheritance ("is a") relationships; lines originating from diamonds indicate "has a" relationships.

The simulation code is implemented in ANSI/ISO C++ using the POOMA Framework [3, 4], and making use of the template facilities of C++. The objects shown in Fig. 1 correspond to C++ classes used in the application. These classes are templated on the number of dimensions and the floating-point type, making it possible to use the same source code base for simulations of different dimensions or data type precision. For the small fraction of the code which cannot be generalized to a dimension-independent formulation, specializations of the relevant functions are provided. At present, this specialization has been done for two and three dimensions.

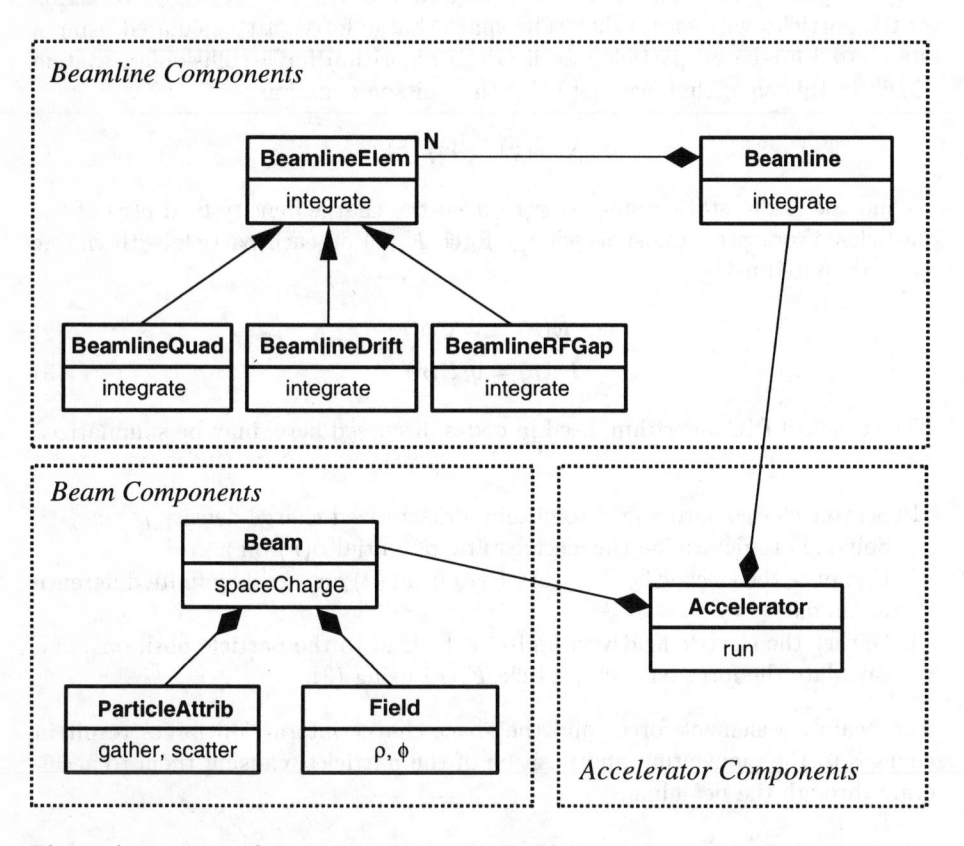

Fig. 1. A summary of the object design for the linear accelerator simulation code. An `Accelerator` consists of a `Beam` (a collection of charged particles) and a `Beamline` (a set of N `BeamlineElems`)

The `Accelerator` class contains the primary components of the simulation, namely a `Beam` instance and a `Beamline` instance. When created, `Accelerator` objects determine simulation parameters and beamline components from an input file, and initialize their `Beam` and `Beamline` accordingly. The `run()` method carries out the steps of the computation, by calling the `integrate()` method of the `Beamline`. The `Beamline` in turn propagates the particles through each individual `BeamlineElem`, which are polymorphic classes that compute specialized forces used to update the momentum and position of the `Beam` particles. The `BeamlineElem` computations invoke the `spaceCharge()` method of the `Beam` to calculate the space-charge interaction forces for the particles.

The accelerator simulation code is built upon the POOMA Framework, a templated C++ class library which provides C++ abstractions for physical quantities such as particles and fields. POOMA provides N-dimensional parallel data structures for the beam particles and for the space-charge field quantities such as the charge density $\rho(r)$, electrostatic potential $\phi(r)$, and electric field $E(r)$. POOMA encapsulates the complexity of providing a parallel run-time system, maintaining parallel data structures, and efficiently performing data-parallel computations. C++ template techniques such as expression templates [7] are used to implement a data-parallel syntax for expressions involving field and particle quantities; such expressions are evaluated at the same speed as hand-coded evaluation loops [3]. POOMA allows the user to write scientific simulation codes that can be run serially or in parallel with no change to the source code. The `Beam` class in Fig. 1 uses POOMA `ParticleAttrib` objects for the particle position and momentum data, and POOMA `Field` objects for $\rho(r)$, $\phi(r)$, etc.

The solution of the Poisson equation from (1) is computed with an FFT-based algorithm that uses multi-dimensional FFT routines from the POOMA Framework. POOMA also provides a number of particle-field interaction capabilities such as gather/scatter algorithms with different interpolation schemes. At present, both cloud-in-cell [8] and nearest-grid-point interpolation mechanisms are supported; additional algorithms are straightforward to implement and use with the POOMA gather/scatter routines.

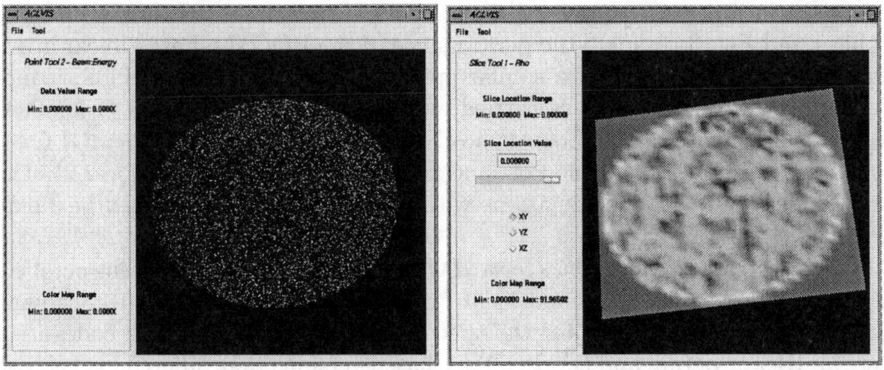

(a) **(b)**

Fig. 2. Visualization of a sample 2D accelerator simulation. (a) Particle positions colored by kinetic energy. (b) Charge density field $\rho(r)$

The POOMA Framework also provides a run-time visualization option that can be used to visualize particle and field data structures either at run-time or post-processed from data files. Figure 2 shows a sample visualization from a 2D linear accelerator simulation, using the POOMA run-time visualization facilities. Figure 2a displays the positions of particles within the accelerator colored by their kinetic energy, and Fig. 2b displays the charge-density field $\rho(r)$ that results from scattering the electric charge of the particles onto a grid.

The use of a toolkit such as the POOMA Framework for development of high-performance simulation codes has proven to be an important tool in the implementation of the linear accelerator simulation code. The strong support in C++ for object-oriented programming features such as polymorphism, inheritance, and data abstraction, coupled with C++'s template facilities, makes it a useful language with which to implement a scientific application such as this. Also, templates provide a mechanism to avoid unnecessary run-time costs normally associated with the use of languages that support OO design, while still maintaining a high degree of flexibility and extensibility in a program.

Software development frameworks such as POOMA have proven to be a powerful tool for high-performance parallel scientific applications. The POOMA Framework has been used for several other codes in fields such as neutron transport [9], and as a basis for other frameworks such as Tecolote [10]. Several other libraries such as PETSc [11], which includes several linear and nonlinear system solvers, and Overture [12], which provides explicit support for overlapping grids in complex geometries, are used as a basis for parallel simulation codes in a wide range of applications. The advantage of using these different systems is clear: building your simulation code on top of an existing parallel application framework simplifies application design, shortens development time, and improves portability to different parallel platforms and communication mechanisms.

4 Performance

Table 1 and Fig. 3 compare the performance of the POOMA-based linear accelerator simulation code with a similar application written in High-Performance Fortran. This comparison was carried out on the Silicon Graphics Origin2000 parallel supercomputers at Los Alamos National Laboratory, using the SGI C++ compiler (version 7.2) and the Portland Group HPF compiler (version 2.2). The calculations were all 2D simulations with a beamline comprising ten beamline elements.

Table 1 shows running times for a 2D fixed-size problem on different numbers of processors. The problem modeled 10^6 particles moving through 10 beamline elements, using a 256^2 grid for the space-charge computation. The codes used were an HPF program and two POOMA-based versions that differed in their use of FFT routines. The POOMA code labeled "C-C" in the table used a complex-to-complex FFT algorithm, and the POOMA code labeled "R-C" used real-to-complex FFT routines. All three codes produced equivalent diagnostic results. The table gives the total simulation time (averaged across the processors) and the amount of time spent in the gather/scatter and FFT portions of the space-charge computation, which is the single largest part of the simulation time.

Table 1. Run times (seconds) for a fixed problem size (10^6 particles, 256^2 grid)

Nodes	Total R-C	C-C	HPF	Gather/Scatter R-C	C-C	HPF	FFT R-C	C-C	HPF
1	537.2	608.6	1998.8	385.6	392.5	1500.2	31.5	83.5	120.1
2	312.0	340.2	1300.8	197.6	198.3	1037.3	23.5	44.7	70.8
4	171.2	184.2	873.2	99.1	99.4	714.9	13.6	24.2	51.3
8	96.9	104.0	467.2	49.3	49.7	384.3	7.4	13.0	24.5
16	61.7	65.0	195.8	24.6	24.7	157.5	4.7	7.4	11.2
32	44.6	46.8	157.1	12.2	12.2	120.8	3.9	5.4	14.4

From the first three columns of Table 1, which list the total simulation time, we see that the POOMA codes outperformed the HPF codes by a factor between

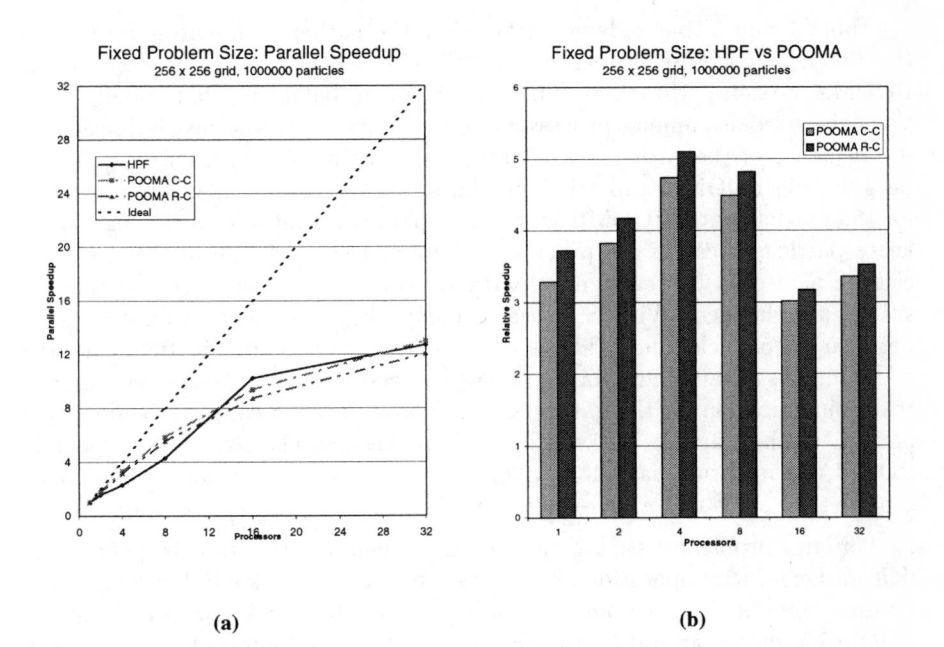

(a)   (b)

Fig. 3. Performance comparison between POOMA and HPF implementations of the accelerator simulation code, in two dimensions. (a) Parallel speedup (single-processor simulation time divided by multi-processor simulation times) for POOMA and HPF codes for a simulation of 10^6 particles on a 256^2 grid. (b) Relative speedup of POOMA over HPF codes for the same problem (HPF simulation time divided by POOMA simulation time)

3 and 5. Figure 3a, which shows the parallel speedup of the three codes, and Fig. 3b, which shows the speedup of the two POOMA codes relative to the HPF code, both demonstrate that the improvement is consistent from one to 32 nodes. The largest improvement between the timings for the HPF code and the POOMA codes is in the time to perform the gather and scatter operations, shown in the middle three columns of Table 1. The times to perform the FFT operations for the POOMA codes were also shorter than for the HPF code, particularly for the real-to-complex version of the POOMA code, but for this problem size the gather/scatter time represents the majority of the computation.

The performance gain with a real-to-complex FFT, which requires less storage and fewer elements in the FFT calculation, is particularly noticeable for problems with a small number of particles per cell. A set of simulations of 10^5 particles on a 256^2 grid using increasing numbers of nodes is summarized in Table 2, using the C-C and R-C versions of the POOMA application. Here, the relative improvement in performance using the real-to-complex version is much more noticeable than in the 10^6-particle simulation. While the parallel speedup for the C-C version is greater than that of the R-C version, the R-C version has much better single-node performance and reaches the point of diminishing parallel returns earlier than the C-C code.

Table 1 and Table 2 demonstrate that the gather and scatter portions of the POOMA codes scale reasonably well with the number of processors. The POOMA version performs an initial particle load-balancing that equally partitions the particles among processors and contributes to the nearly linear scaling behavior of the gather/scatter routines. This highlights a major difference between the POOMA and HPF simulation codes: the parallelization strategy for the particle data. POOMA employs a spatial decomposition strategy, which keeps particles local to the processor containing their charge density field and electric field data by reassigning particles to processors when the particle positions are changed. With a spatial decomposition, gather/scatter operations between the particles and fields require a minimum of communication. The HPF code employs a static partitioning of particles across the processors, requiring extra communication for the gather/scatter phase. In both cases, a roughly equal portion of the particles is kept on each processor. The extra time spent by POOMA to maintain particle locality and to perform the initial load balancing is more than made up for by reduction in the times for gather/scatter operations.

For large problem sizes, the majority of the computation time is spent in particle gather/scatter operations. In addition to the use of a spatial decomposition strategy to minimize the communication during gather and scatter calculations, POOMA provides an option to cache the particle-field interpolation generated in one gather or scatter operation for later gather/scatter calls. Interpolation between particle and field positions involves determination of nearest grid positions and interpolation weights, which do not change from one gather/scatter call to the next unless the particle positions change. For these linac simulation codes, the particles do not move between the time when charge is scattered onto the charge-density field and when the electric field vectors are gathered back to

Table 2. Run times (seconds) for a fixed problem size (10^5 particles, 256^2 grid)

Nodes	Total R-C	Total C-C	Gather/Scatter R-C	Gather/Scatter C-C	FFT R-C	FFT C-C	Speedup R-C	Speedup C-C
1	93.4	158.6	38.8	39.4	31.2	83.4	-	-
2	58.9	84.6	20.0	19.9	23.4	25.3	1.59	1.87
4	32.5	45.9	9.7	9.9	13.3	24.3	2.87	3.46
8	17.8	24.9	4.7	4.8	7.1	12.8	5.25	6.40
16	10.9	14.3	2.3	2.4	4.1	6.8	8.57	11.09
32	8.9	10.2	1.3	1.3	3.5	4.7	10.49	15.55

determine the electrostatic force. By caching the interpolation information from the scatter and reusing it during the gather, the gather operations in the 2D POOMA codes are seen to run up to three times faster than the corresponding scatter operation.

Table 3 compares the execution times for the POOMA and HPF versions of the linac simulation code on two different parallel architectures. In addition to the Origin2000 machines at Los Alamos National Laboratory, the codes were run on the Cray T3E at the National Energy Reseach Scientific Computing Center. On the T3E, the POOMA code was compiled with the Kuck and Associates KCC 3.2b2 compiler (version 3.2d), and the HPF code was compiled with the Portland Group HPF compiler (version 2.4). The results in Table 3 are for a 2D simulation of $500,000$ particles on a 256^2 grid, and the real-to-complex FFT version of the POOMA code was used. On the T3E, the POOMA version executes from just about the same speed to 50 percent faster than the HPF code. This scaling is not as dramatic as what is observed on the Origin2000 machines, but is consistent with the previous results in that the difference in times is due primarily to faster gather/scatter operations in the POOMA implementation.

Table 3. Run times (seconds) for different architectures (500000 particles, 256^2 grid)

Nodes	SGI Origin2000 R-C	SGI Origin2000 HPF	Cray T3E R-C	Cray T3E HPF
1	291.0	1064.7	473.2	586.6
2	170.1	708.0	263.5	370.6
4	113.5	397.5	143.0	198.9
8	65.9	247.2	80.3	110.8
16	38.7	107.3	50.9	63.6
32	31.8	107.6	36.7	35.8

5 Conclusions

Using the POOMA Framework, a C++ application which models the motion of high-intensity charged particle beams through a linear accelerator has been developed that runs substantially faster than an equivalent HPF application on a number of different platforms. This performance increase can be attributed in part to the use of a spatial decomposition strategy for the parallel computation in the POOMA version of the code that reduces the parallel communication required during parallel gather and scatter operations, and in part to the use of a real-to-complex FFT algorithm in the POOMA version. The linac simulation code employs an object-oriented design strategy; by using the POOMA Framework as a basis for the development, the design is able to focus on the specific physics abstractions of the accelerator in a modular, extensible manner. POOMA automatically provides the parallel data structures and algorithms, efficient evaluation of data-parallel expressions, and abstractions of the hardware-specific parallel communication issues for the accelerator code.

References

[1] Robert D. Ryne and Salman Habib. Beam dynamics calculations and particle tracking using massively parallel procesesors. *Part. Accl.*, 55:365, 1996.

[2] Graham A. Mark, William F. Humphrey, Julian C. Cummings, Timothy J. Cleland, Robert D. Ryne, and Salman Habib. Modeling particle accelerators using C++ and the POOMA framework. In *ICNSP '98*, February 1998. Santa Barbara, CA.

[3] William Humphrey, Steve Karmesin, Federico Bassetti, and John Reynders. Optimization of data-parallel field expressions in the POOMA framework. In *ISCOPE '97*, December 1997. Marina del Rey, CA.

[4] John Reynders et al. POOMA: A framework for scientific simulations on parallel architectures. In Gregory V. Wilson and Paul Lu, editors, *Parallel Programming using C++*, pages 553–594. MIT Press, 1996.

[5] C. K. Birdsall, A. B. Langdon, and H. Okuda. Finite-size particle physics applied to plasma simulation. *Methods Comput. Phys.*, 9:241–258, 1970.

[6] A. B. Langdon. Energy conserving plasma simulation algorithms. *J. Comput. Phys.*, 12:247–268, 1973.

[7] Todd Veldhuizen. Expression templates. Technical Report 5, C++ Report 7, June 1995.

[8] C. K. Birdsall and D. Fuss. Clouds-in-clouds, clouds-in-cells physics for many-body plasma simulations. *J. Comput. Phys.*, 3:494–511, 1969.

[9] Stephen Lee, Julian Cummings, and Steven Nolen. MC++: Parallel, portable, monte carlo neutron transport in C++. Technical report, Los Alamos National Laboratory, 1996. Document LA-UR 96-4808.

[10] Jean Marshall, John Hall, Lee Ankeny, Sean Clancy, Jodi Heiken, Kathy Holian, Stephen Lee, Guy McNamara, James Painter, and Mark Zander. Tecolote: An object-oriented framework for physics development. Technical report, Los Alamos National Laboratory, April 1998. Document LA-UR 98-1319.

[11] Lois McInnes and Barry Smith. PETSc 2.0: A case study of using MPI to develop numerical software libraries. In *The MPI Developers Conference*, June 1995. Notre Dame, IN.

[12] David Brown, William Henshaw, and Daniel Quinlan. Overture: An object-oriented framework for solving partial differential equations. In *ISCOPE '97*, December 1997. Marina del Rey, CA.

An Evaluation of Java for Numerical Computing*

Brian Blount and Siddhartha Chatterjee

The University of North Carolina, Chapel Hill, NC, USA

Abstract. We describe the design and implementation of high performance numerical software in Java. Our primary goals are to characterize the performance of object-oriented numerical software written in Java and to investigate whether Java is a suitable language for such endeavors. We have implemented JLAPACK, a subset of the LAPACK library in Java. LAPACK is a high-performance Fortran 77 library used to solve common linear algebra problems. JLAPACK is an object-oriented library using encapsulation, inheritance, and exception handling. It performs within a factor of four of the optimized Fortran version for certain platforms and test cases. When used with the native BLAS library, JLAPACK performs comparably with the Fortran version using the native BLAS library. We conclude that high-performance numerical software could be written in Java if a few concerns about language features and compilation strategies are addressed.

1 Introduction

Java [1] has achieved rapid success due to several key features. Java bytecodes are portable, so programs can be run on any machine that has an implementation of the Java Virtual Machine (JVM). Java provides garbage collection, freeing programmers from concerns about memory management and leaks. The language contains no pointers and dynamically checks array accesses, which help avoid common bugs in C programs. Java is establishing itself as a language of choice for many software developers.

Java is attractive to the scientific computing community for the same reasons. However, several factors limit Java's inroads. First, Java performance has been a source of concern. Many of the attractive features of Java caused early interpreted versions of the JVM to perform poorly when compared with compiled languages like Fortran and C. Second, the absence of a primitive complex type presents another obstacle, as many numeric codes make extensive use of complex numbers. Finally, several language features that make numeric codes less cumbersome to write, such as operator overloading and parametric polymorphism, are absent in Java.

However, we believe that Java may be suitable for writing high-performance numerical software. The problems discussed above can be partially circumvented

---

* This research is supported in part by the National Science Foundation under CAREER award CCR-9501979 and grant CCR-9711438.

by careful programming techniques. Furthermore, certain language features, such as primitive complex types, may be included in future versions of Java. To test our hypothesis that good performance can be achieved in Java, we designed and implemented JLAPACK, a proof-of-concept version of LAPACK in Java. LAPACK is a high-performance Fortran 77 library that solves common linear algebra problems. This library is well-suited for our study for several reasons: it is a standard library in the scientific community; it is used to solve common and useful problems; and it is highly optimized, giving us a hard performance bound.

Our implementation of JLAPACK follows the Fortran version closely in spirit and structure. However, we did not write Fortran-style code in Java. JLAPACK employs object-oriented techniques such as inheritance, dynamic dispatch, and exception handling. We use classes to represent vectors, matrices, and other objects. We use exceptions for error handling. For performance analysis, we ran our code using a fully compliant JVM, with bounds checking and garbage collection enabled. JLAPACK performs within a factor of four of the optimized Fortran version for certain platforms and test cases.

2 LAPACK

LAPACK [2] is a library of Fortran 77 routines for common linear algebra problems, such as systems of linear equations, linear least square problems, eigenvalue problems, and singular value problems. LAPACK uses block-oriented algorithms for many operations, providing more locality of reference and allowing the use of matrix-matrix operations. The library handles both real and complex numbers, with versions for both single and double precision representations. There are specialized routines for structured matrices, such as banded matrices, tridiagonal matrices, and symmetric positive-definite matrices. JLAPACK currently implements only the simple linear equation solver for general matrices (i.e., xGESV and the routines they require) with both blocking and nonblocking versions.

LAPACK uses the Basic Linear Algebra Subroutines (BLAS) [3, 4, 5, 6, 7, 8] for many of its time-critical inner loops. Most high performance machines have BLAS libraries with machine-specific optimizations, called *native BLAS*. Generic Fortran 77 BLAS code is available and is distributed with LAPACK. For JLAPACK, we provided two versions: one implemented in Java, and the other employing vendor-supplied native BLAS. The latter version provides Java wrappers around the Fortran BLAS routines, using the `native` method call mechanism of Java. Bik and Gannon [9] have shown that native methods can be used to achieve good performance, and our findings support their results.

3 JLAPACK

JLAPACK and JBLAS are our Java implementations of the LAPACK and BLAS libraries, currently implementing the subset of the subroutines in both libraries that are used by the *simple general equation solver*. We follow the Fortran version in spirit and in structure, with every Fortran subroutine corresponding to a Java method. We retain the Fortran naming conventions, providing implementations for four data types: single precision real (S), double precision real (D), single precision complex (C), and double precision complex (Z).

Several goals influenced the design of JLAPACK. First, we wanted to encapsulate all the information specifying a vector or matrix into a class. This information fits into two categories that should be kept orthogonal: the data and its shape. Second, we wanted to store matrix data in a one-dimensional array for two reasons: first, two-dimensional arrays in Java are not guaranteed to be contiguous in memory, so a one-dimensional array provides more locality of reference; second, accessing an element in a two-dimensional array requires bounds checks on both indices, doubling bounds checking overhead. Third, we wanted to allow matrices and vectors to share data. A vector object that represents a column of a matrix should be able to use the same data as the matrix itself. Our final goal was to limit the number of constructor calls, as this is a known source of overhead in naive object-oriented programs.

Our design contains three separate components: the JLASTRUCT package, the JBLAS package, and the JLAPACK package. The JLASTRUCT package supplies the vector, matrix, and shape classes used by the library. The JBLAS and JLAPACK packages contains the BLAS library code and LAPACK library code respectively. Both contain four classes, one for each data type. Because there are no instance members in either class, all the methods are static. Each method in the JBLAS classes corresponds to a subroutine in the BLAS library and each method in the JLAPACK classes corresponds to a subroutine in the LAPACK library. We now discuss in detail the design of these packages.

3.1 The Vector, Matrix, and Shape Classes

In Fortran 77, information about the shapes of vectors and matrices must be represented as scalar variables and passed as extra arguments to every routine manipulating vectors and matrices. The vector and matrix classes in our design encapsulate this information into the abstraction of *shape*. There are vector and matrix classes for each of the four data types.

The class JLASTRUCT.Vector implements two methods:
eltAt(i) returns the ith element in the vector
assignAt(val, i) stores *val* in the vector's ith element
The class JLASTRUCT.Matrix implements for matrices:
EltAt(i, j) returns the element at location (i, j)
assignAt(val, i, j) stores *val* at location (i, j)
colAt(i, v) aliases the vector v to the ith column
rowAt(i, v) aliases the vector v to the ith row
submatrix(i, j, r, c, m) aliases the matrix m to the submatrix of size (r, c) starting at location (i, j)

These classes contain two members: *data* and *shape*. The data member is a one-dimensional array of the appropriate type that is guaranteed to contain all the vector/matrix elements. The shape member is of type JLASTRUCT.VShape (for vectors) or JLASTRUCT.MShape (for matrices), both classes being subclasses of the abstract class JLASTRUCT.Shape. The shape object defines the layout of the vector or matrix elements in the data array.

An object of type JLASTRUCT.VShape contains the members:
start: The index in *data* of the first vector element
len: The number of vector elements
inc: The step size in *data* between consecutive vector elements

Therefore, element i of a vector resides in slot j of its data array, where $j = start + i * inc$. Elements of a vector are evenly spaced in the data array.

An object of type JLASTRUCT.MShape contains the members: *start:* The index in *data* of the first matrix element
rows: The number of rows in the matrix
cols: The number of columns in the matrix
ld: The distance in *data* between the first elements in consecutive columns

Therefore, matrix element (i, j) resides in location k of its data array, where $k = start + ld * j + i$. Note the column-major storage order and the zero-based indexing of arrays. This fits the Fortran model, allowing JLAPACK to use the same optimizations as the Fortran version and enabling native BLAS to be incorporated.

This implementation allows objects to share data arrays. Figure 1 shows how this may occur. The ability to share member objects improves the performance of methods used to obtain rows, columns, and sub-matrices of matrices. We will use the *colAt()* method as an example, as its implementation applies to the other two. A naive implementation of this method would allocate new memory for the vector and new memory for its shape. Instead, the *colAt()* takes as a parameter a vector that has already been allocated. Then, the method only supplies the vector's data member (by giving it a reference to its own data), and updates its shape object. This approach eliminates unnecessary data copying and allows reuse of storage for temporary vectors and matrices.

Boisvert *et al.* [10] discuss an implementation for numerical libraries in Java that does not encapsulate vectors and matrices. They use two-dimensional arrays to represent matrices, and store information describing the shape of vectors and matrices in local variables, similar to the Fortran version. This approach requires several versions of each vector operation. One version must handle the case where a vector is stored in a one-dimensional array, and another must handle the case where a vector is a column of a matrix, and is stored in a two-dimensional array. They claim [10, p. 41]: "If we are to provide the same level of functionality as the Fortran and C BLAS then we must provide *several* versions of each vector operation." While this may be true of *implementations* of BLAS primitives, this should not affect the *interface* visible to the programmer. Our shape abstraction unifies and encapsulates these various cases. For efficiency, an implementation can still provide specialized routines for common cases.

3.2 Limiting Constructor Calls

Excessive object creation is a well-known source of performance loss in object-oriented programs. Therefore, we use a technique (similar to that described by Dingle and Hildebrandt [11]) to limit the number of temporary vector and matrices created. Such objects are used locally in methods of the JBLAS and JLAPACK classes, so it is natural to place them within these methods. However, we make them private static class members. Note that this approach works only because none of the methods in the library are recursive and because we are ignoring issues of thread safety.

	0	1	2	3	4	5	6	7	8	9	10	11	12	13	14	15
data:	1.2	3.4	6.7	2.6	9.4	8.6	2.7	7.3	5.5	9.9	6.7	3.6	3.9	2.1	4.3	1.9

Operation:	**Shape:**	**Object:**
A = new DMatrix(data, 4, 4);	A.data = data	
	A.shape.start = 0	
	A.shape.rows = 4	
	A.shape.cols = 4	
	A.shape.ld = 4	
B = new DMatrix();	B.data = data	
A.submatrix(1, 1, 2, 2, B);	B.shape.start = 5	
	B.shape.rows = 2	
	B.shape.cols = 2	
	B.shape.ld = 4	
C = new DVector();	C.data = data	
B.rowAt(1, C);	C.shape.start = 6	
	C.shape.inc = 4	
	C.shape.len = 2	

Object for A:

1.2	9.4	5.5	3.9
3.4	8.6	9.9	2.1
6.7	2.7	6.7	4.3
2.6	7.3	3.6	1.9

Object for B:

8.6	9.9
2.7	6.7

Object for C:

2.7	6.7

Fig. 1. Sharing of data among multiple matrices and vectors. The 4×4 matrix A uses all 16 elements of its data array. Matrix B is assigned to be a submatrix of A. It shares the same data object as A, but only uses 4 elements of the array. Vector C represents one row of matrix B. Again, it shares the data object with A and B, but only uses two elements. Note how the shape parameters specify exactly where the data is stored.

3.3 Method Granularity

Most of the work in the BLAS routines involves looping through columns of a matrix, accessing and modifying elements. An example is the `scale` routine, which scales a vector by a constant factor. A natural implementation would perform an *eltAt*() and an *assignAt*() call for each element in the vector. Unfortunately, every call to *eltAt*() and *assignAt*() must use the shape object to calculate the address of an element. The vector and matrix access equations above show the cost of these calculations. Boisvert *et al.* [10] observe that the use of such methods is five times slower than an ordinary array access. We employ two mechanisms to overcome this overhead: aggregate operations and incremental access methods.

Aggregate operations are operations performed on an entire vector or matrix at once. We converted operations such as the scale operation into methods in the vector and matrix classes. These methods exploit the bulk nature of the updates to access successive elements using incremental address computations. The calculation of the index into the data array consists only of an increment, instead of the multiplication and addition performed in the *eltAt*() method.

Another common type of operation in the library is to loop over a vector, accessing but not modifying its elements. Because the elements are being used instead of being modified, aggregate methods do not apply. To limit the number of index calculations, we include *incremental access methods*. These methods are used to retrieve the next element of a vector or the next column of a matrix, and are similar to the methods defined by the `java.lang.Enumeration` [12] interface. However, `Enumeration` does not handle primitive types, so we could not implement this functionality with the `Enumeration` interface.

3.4 Complex Numbers

Currently, Java does not provide a primitive type for complex numbers. However, complex numbers are required within the LAPACK library, so we provide two implementations for them. The first approach is to use a class `JLASTRUCT.Complex`, encapsulating complex values and arithmetic operations on them. While this is object-oriented, the overhead of using many small objects and calling a method for every arithmetic operation makes this approach unusably slow.

Our second implementation of complex numbers simply inlines them, by making the data arrays of the vector and matrix classes twice as long, and storing the real and imaginary components contiguously in the array. Access methods change from *eltAt*() to *realAt*() and *imgAt*(), and all arithmetic is performed inline. While this is an unattractive approach to dealing with complex numbers, it demonstrates the performance achievable with a primitive complex type.

3.5 Discussion

We discuss certain aspects of Java that make the development of JLAPACK difficult, and how we address them.

Two language issues hinder the development of JLAPACK: the absence of parametric polymorphism and the absence of operator overloading. The absence of parametric polymorphism required us to create a version of the JLAPACK library for each data type, which results in code bloat and extra programmer effort. Several projects [13, 14, 15] have examined methods for providing parametric polymorphism, either by modifying the JVM or by a adding a preprocessing phase, and it is possible that the feature will be available in future versions of Java.

The lack of operator overloading required us to write many methods in unnatural forms. For example, the *colAt*() method intuitively should return a Vector object. Because we could not overload the assignment operator, we had to pass in the Vector object as a parameter to the method. Likewise, we had to write out in full detail mathematical operations such as scaling of vectors, instead of using a more natural mnemonic form, such as the *= operator.

It is true that neither of these language features is fundamental, and that both represent "syntactic sugar" that would be removed in a preprocessing step. We ignored these issues while implementing JLAPACK, as our goal was to test our hypothesis about performance. However, the general user does not want to deal with such issues and is less apt to use a library that has such unnatural syntax. (Witness the success of Matlab, which virtually removes the difference between the linear algebraic representation of an algorithm and its realization in code.) We feel that Java will not be attractive to the numerical computing community until these features are integrated into the language.

Our results document the overhead of encapsulating complex numbers in classes. Manual inlining is not the correct solution either, as it detracts from the readability of the code, replicates common operations, and presents a common source of bugs. While it is beyond the scope of this paper to determine the best mechanism for including primitive complex numbers in Java, this issue is under consideration by the Java Grande Forum [16], and must be resolved satisfactorily if Java is to be viable for numerical computing.

4 Performance

Performance is an overarching concern for scientific computation. The Fortran version of LAPACK has been highly optimized and represents our target level of performance. Therefore, we compare JLAPACK with the optimized Fortran version (compiled on the test platform with the vendor's optimizing Fortran77 compiler from LAPACK 2.0 source distribution downloaded from www.netlib.org) in all our results. In this section, we present the results from our experiments and discuss the reasons for both good and poor performance.

We present performance results for solving the system of linear equations $AX = B$, using a coefficient matrix A and a right hand side matrix B whose entries are generated using a pseudorandom number generator from a uniform distribution in the range $[0, 1]$. The same seeds are used in both the Fortran and Java versions, to guarantee that both versions solve identical problems. The square matrix A has between 10 and 1000 columns. The matrix B has from 1 to 50 columns. In every case, the leading dimension of the matrix equals the number of rows of the matrix. We separately timed the triangular factorization (xGETRF) and triangular solution (xGETRS) stages. The two data types used in timing were double precision real numbers (x=D) and double precision complex numbers (x=Z). For the factorization stage, we used block sizes between 1 and 64.

Table 1 lists the platforms we used for timing. We ran Fortran versions for all Unix platforms, using the -fast option when compiling the Fortran library. On the DEC, where native BLAS libraries were available through the dxml library [17], we measured performance with both the JBLAS classes and the native library. On the Sparcs, we ran two versions with kaffe [18, 19]: one with dynamic array bounds checking turned on and the other with this feature turned off. We turned off array bounds checking in kaffe by modifying the native instructions that its JIT compiler emits. We measured performance without array bounds checking for two reasons. First, we wanted to quantify the cost of performing bounds checks. Second, global analysis of our code could prove that instances of java.lang.ArrayIndexOutOfBoundsException could never be thrown. While this cannot always be determined from the structure of the program, and no current implementation of the JVM systematically eliminates runtime bounds checking in this manner, such an optimization is likely to appear in future generations of JVM implementations.

We manually compensated for certain deficiencies in javac to boost the performance of our code. The primary modification was loop unrolling. In our experiments, an unrolling depth of four gave the best performance. Unrolling does introduce a cost in code size. Unrolling loops in the JLASTRUCT.Vector class by factors of two, four, and eight increased class file sizes by 41%, 62%, and 104%.

Table 2(a)–(d) presents performance results for the three platforms listed in Table 1, including the test cases on the DEC using the native BLAS library. Analysis of the results reveals several interesting facts. First, the Java version with bounds checking enabled and inlined complex numbers performs within

Table 1. Testing environment

	SparcStation 5	UltraSparc 17	DEC Personal Workstation
Processor	SPARC	Ultra 1	Alpha 21164
Processor Speed	110 MHz	170 MHz	500 MHz
Memory	40 MB	64 MB	512 MB
Operating System	Solaris 2.5.1	Solaris 2.5.1	Digital Unix 4.0D
JVM	Kaffe v0.9.2	Kaffe v0.9.2	JDK 1.1.4
JIT Enabled	Yes	Yes	Yes
F77 Compiler Switches	-fast	-fast	-fast

Table 2. Performance results for double precision real (D) and double precision complex (Z) values. Entries represent the ratio of the JLAPACK running time to the LAPACK running time (lower is better). Results for the complex version that uses inlined complex numbers are denoted by (I), and results for the version that used classes for complex numbers are denoted by (C). The results for the triangular factorization without blocking are denoted by F(nb), the results for the triangular factorization with a blocking factor of 16 are denoted by F(b), and the results for the solve are denoted by S. The label bc denotes that bounds checking was enabled, and nbc denotes that it was disabled. The label r indicates a small matrix (100 by 100) was used so that the program could take advantage of caching. The label R indicates a large matrix (600 by 600) that could not fit into the system cache was used. A — label denotes a missing entry. (a) Performance on a SPARCstation 5. (b) Performance on an UltraSparc 17. (c) Performance on a DEC Personal Workstation. (d) Performance of Native BLAS on a DEC Personal Workstation.

		D		Z(I)		Z(C)	
		r	R	r	R	r	R
F	bc	4.85	3.02	5.27	3.27	9.84	8.42
(nb)	nbc	4.10	2.67	3.87	2.96	—	—
F	bc	5.26	2.57	4.73	3.32	10.69	9.67
(b)	nbc	4.41	2.24	4.49	3.01	—	—
	bc	5.63	2.73	5.61	4.34	13.64	10.47
S	nbc	3.20	2.39	4.50	3.29	—	—

(a)

		D		Z(I)		Z(C)	
		r	R	r	R	r	R
F	bc	11.22	5.23	12.90	6.05	28.68	16.31
(nb)	nbc	9.32	4.17	10.81	5.06	—	—
F	bc	14.33	10.81	8.69	6.24	19.06	19.16
(b)	nbc	12.16	8.69	7.48	5.21	—	—
	bc	16.64	5.43	5.71	3.72	17.33	10.61
S	nbc	11.57	4.45	4.53	3.15	—	—

(b)

		D		Z(I)		Z(C)	
		r	R	r	R	r	R
F	bc	6.49	2.83	6.99	2.73	49.83	19.85
(nb)	nbc	—	—	—	—	—	—
F	bc	8.37	6.22	8.10	6.30	52.82	48.09
(b)	nbc	—	—	—	—	—	—
	bc	4.78	2.82	5.04	3.53	33.65	26.31
S	nbc	—	—	—	—	—	—

(c)

		D		Z(I)		Z(C)	
		r	R	r	R	r	R
F	bc	2.81	0.84	—	—	—	—
(nb)	nbc	—	—	—	—	—	—
F	bc	4.10	1.07	—	—	—	—
(b)	nbc	—	—	—	—	—	—
	bc	1.71	1.15	—	—	—	—
S	nbc	—	—	—	—	—	—

(d)

a factor of four of the Fortran version for certain architectures and problem sizes. On the SparcStation 5, the Java version is about three or four times worse than the Fortran on the larger problem sizes for both the factorization and the triangular solve. As a side note, the interpreted Java implementation was unusably slow.

Second, on the UltraSparc, for most of the cases with bounds checking enabled and inlined complex numbers, there is less than a factor of seven difference between the two versions. However, for the factorization with double precision numbers and blocking, the Fortran version performs about eleven times better than the Java version. This is because blocking significantly improves the performance of the Fortran version, but not of the Java version. Our hypothesis is that the variations in performance represent instruction scheduling effects. We examined the assembly code generated by the Fortran compiler on the Sparc-

Station 5 and on the UltraSparc, which represent different implementations of the same instruction set architecture. The code generated for the inner loops of several routines varied considerably, using different degrees of loop unrolling and different schedules. The `kaffe` JIT compiler generated identical instruction sequences for both platforms. We believe that the sub-optimal instruction schedule increases pipeline stalls and nullifies the improvements in spatial locality due to blocking.

Third, the native BLAS library made a significant impact on performance, especially for the cases where blocking was used. Because LAPACK heavily relies on BLAS for its computations, using the native BLAS library brought the performance of JLAPACK close to the performance of LAPACK (within 15% for large problem sizes). This demonstrates that the object-oriented wrappers provided by JLAPACK were efficient. It also supports our hypothesis that poor instruction scheduling hurt performance in the pure Java version.

Fourth, the impact of bounds checking is shown by the data generated on the Sparcs. For the test cases, removing bounds checking increased performance by 10% to 25%. The affect was slightly larger for the UltraSparc than the Sparc-Station, and slightly larger for the solution stage than the factorization stage.

Finally, using classes to represent complex numbers performs very poorly. On all the platforms tested, the version that uses the `Complex` class is more than twice as slow as the version that inlined complex numbers.

5 Related Work

Several other projects investigate Java for numerical computing. The Java Numerical Toolkit [10] is a set of libraries for numerical computing in Java. Its initial version contains functionality such as elementary matrix and vector operations, matrix factorization, and the solution of linear systems. HPJava [20] is an extension to Java, that allows parallel programming. HPJava is somewhat similar to HPF and is designed for SPMD programming.

Several projects are developing optimizers for Java. Moreira *et al.* [21] are developing a static compiler that optimizes array bounds checks and null pointer checks within loops. Adl-Tabatabai *et al.* [22] have developed a JIT compiler that performs a set of optimizations, including subexpression elimination, register allocation, and the elimination of array bounds checking. Such optimizations may allow us to bridge the performance gap between our version with bounds checking and our version without bounds checking.

6 Conclusions and Future Work

Portability, security, and ease of use make Java an attractive programming environment for software development. Performance problems and the absence of several language features have hindered its use in high-performance numerical computing. While operator overloading and parametric polymorphism are indeed "syntactic sugar", they will contribute significantly to the usability of the language and to the willingness of the numerical computing community to use Java. We have quantified the difference between using a primitive type for complex numbers, which we have simulated, and using a class for complex numbers. As expected, there is strong evidence that a primitive type is needed.

Future work in the development of high-performance object-oriented numerical libraries in Java can be divided into the following categories.

Programming model changes. The algorithms implemented in most numerical libraries today were designed for the Fortran programming model. These may not be the best algorithms when run under the object model of Java. We have discussed several object-oriented programming idioms to implement numerical libraries efficiently. Future work needs to explore these and other techniques such as expression templates [23].

Compiler changes. We noted in Section 4 several desirable optimizations that javac does not perform. Much work remains to be done here to develop better compilation techniques for Java. Budimlic and Kennedy [24] are exploring such optimizations using object inlining techniques.

Just-In-Time compilation. Current JIT compilers are in their early version, and have not been heavily optimized. As we discussed in Section 4, some do not take advantage of machine-specific optimizations and do not appear to schedule code effectively.

Architectural issues. Current trends in processor implementation adds significant instruction re-ordering capabilities to the hardware. Engler [25] conjectures that this may reduce or obviate the need for instruction scheduling by JIT compilers. This is a reasonable conjecture that needs to be tested.

Experimentation with other codes. LAPACK is obviously not representative of all numerical software. Further work needs to be done to determine if Java implementations of other numerical software behave similarly.

Our results show that Java may perform well enough to be used for numerical computing, if a handful of concerns about language features and compilation strategies are adequately addressed. While we have not yet met the goal of having Java perform as well as Fortran, we are beginning to get reasonably close. We speculate that a combination of techniques will narrow this gap considerably over the next few years, and that Java will be the language of choice for numerical computing by the year 2000.

References

[1] K. Arnold and J. Gosling. *The Java^{TM} Programming Language*. The Java^{TM} Series. Addison-Wesley Publishing Company, 1996.

[2] E. Anderson *et al. LAPACK User's Guide*. SIAM, second edition, 1995.

[3] J. J. Dongarra *et al.* A set of level 3 basic linear algebra subprograms. *ACM Trans. Math. Softw.*, 16(1):1–17, Mar. 1990.

[4] J. J. Dongarra *et al.* Algorithm 679: A set of level 3 basic linear algebra subprograms: Model implementaton and test programs. *ACM Trans. Math. Softw.*, 16(1):18–28, Mar. 1990.

[5] J. J. Dongarra *et al.* An extended set of basic linear algebra subprograms. *ACM Trans. Math. Softw.*, 14(1):1–17, Mar. 1988.

[6] J. J. Dongarra *et al.* Algorithm 656: An extended set of basic linear algebra subprograms: Model implementaton and test programs. *ACM Trans. Math. Softw.*, 14(1):18–32, Mar. 1988.

[7] C. L. Lawson *et al.* Basic linear algebra subprograms for Fortran usage. *ACM Trans. Math. Softw.*, 5(3):308–323, Sep. 1979.

[8] C. L. Lawson *et al.* Algorithm 539: Basic linear algebra subprograms for Fortran usage. *ACM Trans. Math. Softw.*, 5(3):324–325, Sep. 1979.

[9] A. J. C. Bik and D. B. Gannon. A note on native level 1 BLAS in Java. *Concurrency: Practice and Experience*, 9(11):1091–1099, Nov. 1997.

[10] R. F. Boisvert *et al.* Developing numerical libraries in Java. In *Proc. ACM 1998 Workshop on Java for High Performance Network Computing*, pages 35–44, 1998.

[11] A. Dingle and T. H. Hildebrandt. Improving C++ performance using temporaries. *Computer*, pages 31–41, Mar. 1998.

[12] D. Flanagan. *Java In a Nutshell*. O'Reilly & Associates, Inc., 1997.

[13] O. Agesin, S. Freund, and J. Mitchell. Adding type parameterization to the Java language. In *Proc. OOPSLA'97*, pages 49–65, 1997.

[14] A. C. Myers, J. A. Bank, and B. Liskov. Parameterized types for Java. In *Proc. POPL'97*, pages 132–145, Jan. 1997.

[15] M. Odersky and P. Wadler. Pizza into Java: Translating theory into practice. In *Proc. POPL'97*, pages 146–159, Jan. 1997.

[16] Java Grande Forum. The Java Grande Forum charter document. http://www.npac.syr.edu/javagrande/jgfcharter.html.

[17] DIGITAL Extended Math Library. http://www.digital.com/hpc/software/dxml.html.

[18] T. J. Wilkinson. The Kaffe homepage. http://www.transvirtual.com/kaffe.html.

[19] M. Barr and J. Steinhorn. Kaffe, anyone? Implementing a Java Virtual Machine. *Embedded Systems Programming*, pages 34–46, Feb. 1998.

[20] G. Zhang *et al.* Considerations in HPJava language design and implementation. In *Proc. LCPC'98*, 1998. To appear.

[21] J. E. Moreira, S. P. Midkiff, and M. Gupta. From flop to megaflops: Java for technical computing. In *Proc. LCPC'98*, 1998. To appear.

[22] A.-R. Adl-Tabatabai *et al.* Fast, effective code generation in a just-in-time Java compiler. In *Proc. PLDI'98*, pages 280–290, Montreal, Canada, June 1998.

[23] F. Bassetti *et al.* A comparison of performance-enhancing strategies for parallel numerical object-oriented frameworks. In *Proc. ISCOPE'97*, 1997.

[24] Z. Budimlic and K. Kennedy. Optimizing Java: Theory and practice. *Concurrency: Practice and Experience*, 9(6):445–463, June 1997.

[25] D. R. Engler. VCODE: A retargetable, extensible, very fast dynamic code generation system. In *Proc. PLDI'96*, pages 160–170, 1996.

High-Level Parallel Programming of an Adaptive Mesh Application Using the Illinois Concert System

Bishwaroop Ganguly* and Andrew Chien**

University of Illinois, Urbana, Illinois, USA

Abstract. We have used the Illinois Concert C++ system (which supports dynamic, object-based parallelism) to parallelize a flexible adaptive mesh refinement code for the Cosmology NSF Grand Challenge. Out goal is to enable programmers of large-scale numerical applications to build complex applications with irregular structure using a high-level interface. The key elements are an aggressive optimizing compiler and runtime system support that harnesses the performance of the SGI-Cray Origin 2000 shared memory architecture. We have developed a configurable runtime system and a flexible Structured Adaptive Mesh Refinement (SAMR) application that runs with good performance. We describe the programming of SAMR using the Illinois Concert System, which is a concurrent object-oriented parallel programming interface, documenting the modest parallelization effort. We obtain good performance of up to 24.4 speedup on 32 processors of the Origin 2000. We also present results addressing the effect of virtual machine configuration and parallel grain size on performance. Our study characterizes the SAMR application and how our programming system design assists in parallelizing dynamic codes using high-level programming.

1 Introduction

The challenges of parallel programming include load balancing, data distribution, and coordination of communication between separate streams of execution. In general, all three of these issues must be addressed in order to obtain the greatest performance benefit. Hardware cache-coherent shared memory multiprocessors offer a shared virtual address space with caches that are kept coherent automatically by hardware. This supports parallel programming by reducing the latency of remote memory access, making communication costs less critical. In practice, inter-processor communication costs remain a key factor in determining performance. The latency of remote memory access due to cache misses is still high when compared to processor clock cycles and can greatly hamper parallel efficiency.

---

* Research scientist at Massachusetts Institute of Technology Lincoln Laboratory
** Science Applications International Chair Professor at UC, San Diego

Shared memory multiprocessors are being scaled up to 512 or even 1024 processors. Studies are needed to determine what runtime system support can utilize this trend and what applications are enabled as a result. Structured Adaptive Mesh Refinement (SAMR) methods are an important class of numerical codes and are a target for such investigations because they contain dynamic parallelism and are difficult to program using a low-level, message passing programming paradigm.

Our research addresses the use of large-scale shared memory systems, utilizing a highly efficient runtime system to support application parallelization. Our runtime system allows programs to utilize varying degrees of shared memory support. To demonstrate our technology, we have parallelized a large-scale SAMR application, a numerical simulation method being used for the NSF Cosmology Grand Challenge [1].

SAMR is a technique for simulating a discrete model of space. It works by solving hyperbolic partial differential equations numerically through a series of discrete timesteps. SAMR methods recognize that a large portion of this modeling area is often empty and/or constant throughout the simulation. SAMR uses a dynamic hierarchy of meshes, only creating high-resolution meshes where heuristics deem that they are needed. This approach saves computation while still providing high resolution modeling. SAMR has been successfully applied to a number of important problems [1], and is gaining acceptance in the realm of scientific computation. Programming SAMR is more complex than single or multi-grid simulations, and this has prevented it from becoming more widely used. Data structures (i.e. meshes) in SAMR are created and deleted from the hierarchy as appropriate to achieve desired solution precision. Therefore, the parallelism across meshes is *dynamic*, and is not analyzable during compilation or even initialization of the program. SAMR's dynamic parallelism must be exploited during the run of the program, making parallelization of the method a challenge.

The Illinois Concert System [2] and ICC++ language [3, 4] together form a parallel programming environment geared towards tackling dynamically parallel codes. This system provides a variety of support for dynamic parallelism and distributed parallel data structures – a global namespace, efficient fine-grained threads, orthogonal data distribution, and high performance runtime primitives. These capabilities make it possible to express dynamically parallel applications with modest effort, and preserves program flexibility for application tuning or algorithm improvement. The ICC++ parallel language is based on the object-oriented language C++. It provides simple syntactic language extensions to express parallel blocks and loops in code, and a distributed data type for concurrent object distribution.

This paper is a case study for parallelizing a dynamic application on large-scaled shared memory using a high-level object-oriented programming environment. The contributions of this work are to show high absolute performance using a high-level parallel programming system, and to explore configurations of our runtime system, gauging how well they exploit the underlying hardware.

In our SAMR implementation the meshes are *tiled*, or partitioned into independent pieces in order to control granularity of parallelism. Our performance results show that a modest parallel grain size of 50x50 tiled meshes is best to provide ample parallelism, and minimize overhead of concurrent execution. We achieve up to 24.4 times speedup on a 32 processor Origin 2000 [5] using maximum shared memory support for communication, which is 93% of maximum feasible speedup for our application. In addition, we achieve 11.7 times speedup on 16 processors for our application by relying heavily on our high-performance messaging support and simple additional techniques. Both of these performance results are promising, and demonstrate the usefulness of our configurable runtime system on shared memory machines. The good performance shows that high-level parallel programming with the Illinois Concert System is viable and can aid greatly in capture of dynamically parallel applications.

The rest of this paper is structured as follows. Section 2 provides background on concurrent object-oriented programming and SAMR methods. Section 3 describes our SAMR implementation in detail. Section 4 discusses the Concert system, with particular attention to runtime system capabilities. Section 5 shows detailed performance results, and several conclusions that we have drawn. Section 6 highlights some related work. Section 7 summarizes our study, and briefly discusses future work.

2 Background

Concurrent Object-Oriented Programming (COOP) builds on the idea of objects for encapsulation in sequential programs, extending it for concurrent programs. Objects in COOP are seen as autonomous communicating entities. This view maintains the benefits of sequential object-orientation by providing reuse and modularity. It also allows for flexible concurrency, where objects can be distributed and operated upon either explicitly by the programmer or automatically by a compiler or runtime system.

The Illinois Concert System is a programming environment that harnesses the benefits of COOP, with a goal of high performance. It consists of the ICC++ language [4], Concert compiler [2, 6, 7] and the Concert runtime system [8]. Concert supports fine-grained, concurrent object-oriented programming on Actors [9]. Computation is expressed as method invocations on objects or collections of objects. Concurrent method invocations operate against state stored in dynamically created thread data structures. Synchronization of concurrent threads is handled automatically by the system, and the user only needs to be concerned with specifying concurrent parts of the code and not concurrency management. ICC++'s syntactic similarity to C++ provides ease of annotation of sequential C++ applications for concurrency.

In addition to common features of sequential object-oriented languages such as object encapsulation and inheritance, three features of the programming model support programming dynamic parallel applications. *A shared name space* allows programmers to build sophisticated distributed data structures without explicit name management. *Implicit dynamic thread creation* frees programmers from explicit thread and synchronization management. *Object-level concurrency control* maintains sequential consistency of the object state in the global namespace, freeing the programmer from managing explicit locking.

The Concert system implementation [2, 6, 7, 4, 10] is a state-of-the-art implementation of a COOP programming model. A full discussion of the Concert runtime system can be found in [10]. The Concert compiler [4] implements a number of aggressive, inter-procedural optimizations that achieve high performance for sequential object-oriented codes. An implementation of Fast Messages [10] is used for low-overhead communication between address spaces.

Adaptive mesh methods are a class of finite difference method that provide high modeling resolution and computational efficiency by generating a hierarchy of grids, comprised of a series of levels. Each level contains a list of grids, with grids on a lower level modeling progressively smaller physical areas of space. Grids at lower levels derived from higher-level grids are said to be *subgrids* or *child grids*, with the higher level grid acting as the child grids *parent*. Furthermore, the hierarchy is dynamic in that grids can be created and deleted from the hierarchy as the simulation proceeds. This framework sets the stage for adaptive methods and provides the benefits of computational efficiency as well as arbitrary resolution.

3 SAMR Implementation

Our study is based on a sequential SAMR method written by scientists at NCSA in the Computational Cosmology group. It has been used in a number of cosmology experiments [1]. It is written in C++ and uses Fortran 77 kernels to perform the compute-intensive interpolation and partial differential equation solves.

3.1 Code Structure

The code is structured as a series of phases. Each phase is either an inter-mesh communication phase or is mesh independent. An example of a mesh independent phase is the partial differential equation solve, and an example of a communication phase is the mesh-to-mesh boundary copy phase. The code first generates an initial hierarchy of meshes to model the physical space, and then iterates over the phases, for each timestep dt, at each level of the hierarchy.

After each iteration of the phases, the entire hierarchy is regridded from the current level down. The first three of the seven phases are mesh independent. The rest of the phases (including the regridding) involve inter-mesh communication. Hierarchies generated by our code typically consist of hundreds of 2D grids each ranging in size from 100 to 1000 cellpoints in each dimension.

3.2 Code Parallelization

Each of the phases listed above contain parallelism over meshes, or pairs of meshes at a level. Dependences in our particular implementation force us to be able to exploit only intra-level parallelism. Even so, almost the entire method can be parallelized.

Current popular parallelization approaches struggle to capture parallelism over dynamic data structures created during the run of a program. Explicit message passing systems such as MPI force users to manually track the location of every dynamically created data structure. This is not a good match for the SAMR method, which creates an entirely new hierarchy of possibly hundreds of new grids at every iteration. Compiler-based approaches such as HPF [11] need compile-time knowledge to load balance such an application, and are usually unable to glean such information through analysis. We have also explored the use of thread-based shared memory to parallelize the code. In brief, these efforts showed that the support that exists on shared memory machines lack the flexibility and robustness to parallelize codes that are as fine-grained as our SAMR implementation, without a great deal of additional programming effort.

ICC++ and Concert allow the user to annotate dynamic concurrency in a clear, simple way and manages the concurrent execution automatically. By making each mesh an object, we express each parallel phase as a series of concurrent method calls as follows:

```
class Grid {
....
public:
void Solve();
};
conc for(all grids G at this level)  // Solve loop
  G.Solve();
```

The ICC++ conc annotation in the for loop above annotates the loop as concurrent. All of the calls to the Solve routine execute independently, and synchronize barrier-style at the end of the loop.

3.3 Tiling

There are inherent problems with a simplistic approach of distributing meshes for concurrency. For load balance, mesh-based distribution depends on the algorithm to provide parallelism. For example, if the method's heuristics choose to create only three grids at a particular level, and we are running on four processors, there is no way to achieve good load balance.

To address this problem, we have employed a technique called *tiling*. In tiling, we use a source code transformation to partition each mesh into equal-sized parts. Our original grids become distributed arrays of tiles, and the tiles are distributed among processors. We have ensured that this transformation is transparent, and produces the same results as the untiled simulation. Tiling gives us controllable, uniform granularity, so that data structures can be distributed evenly across address spaces. The cost of the transformation is increased communication, as each tile has its own boundary region. We shall see that for certain tile sizes, the load balance benefits of tiling far outweigh the additional communication cost incurred in terms of parallel performance.

4 Concert Runtime Configuration

This section will describe the Concert runtime system configurations on shared-memory machines. The Concert runtime logically distributes the underlying memory into *address spaces* over which a program's objects are distributed for concurrent execution. The address spaces each contain heavyweight threads (e.g. Unix processes) to execute lightweight logical Concert threads. A full description of the system can be found in [10]. We have developed a configurable version of this system for the shared memory architecture.

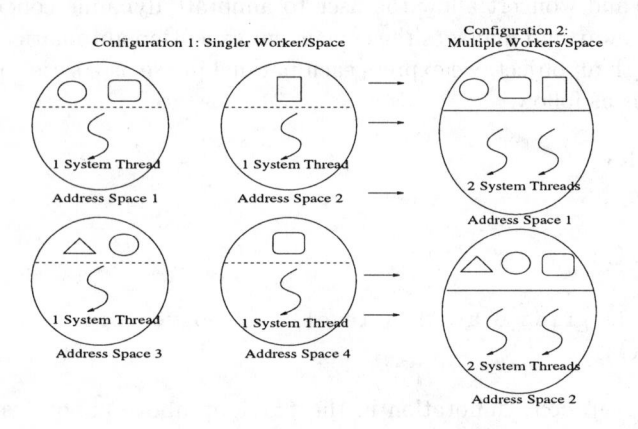

Fig. 1. Runtime configurations

The Concert runtime implementation on the shared memory SGI-Cray Origin 2000 machine is architected to take advantage of the hardware shared memory support. The system is configurable to utilize a range of levels of shared memory support versus messaging support for communication. This is accomplished with *multiple* system threads in each address space. These threads share a work queue (see Fig. 1). Assuming an even distribution of objects, address spaces in Configuration 1 tend to contain fewer objects each than in Configuration 2 for a given number of object instances in a program. Thus sharing in Configuration 2 utilizes more shared memory support, at a cost of decreased locality and processor-data affinity. This flexibility in design allows applications to choose a degree of utilization of shared memory support. We can enumerate a range of configurations, including pure shared memory, hybrid models with multiple thread per address space, and pure distributed memory. In general, these configurations are listed from least to most messaging support use, and from least to most processor-data affinity.

5 Performance Results

Our investigation evaluates the use of high-level programming to get performance from the target application. Related to this, we identify several issues that we will explore with our performance experiments. *Granularity Versus Available Parallelism*; the tiling transformation allows us to tune our application to run with a range of parallel grain sizes and available parallelism. We vary the tiling parameter and measure the effects on performance. *Runtime Configuration* allows us to compare the benefits of shared memory utilization to those of increased data locality. The overall goal is to evaluate system performance by varying the way in which we utilize the underlying shared memory hardware. We also seek a high level of absolute performance for the SAMR code. All experiments were run on the NCSA Origin 2000, on configurations of up to 32 processors with 12Gb of total physical memory running IRIX version 6.4.

The input data set for our experiments is a two-dimensional ShockTube simulation [1]. This test case can be run either adaptively, or as a non-adaptive, single-mesh simulation. Size of the input data set is roughly governed by the size of the top-level mesh in the hierarchy, which is specified as a program parameter. For a complete description of test cases and a full set of results, see [12].

We benchmarked the ICC++ code against the C++ compiled using the SGI CC compiler with full (-O3) optimization and found the ICC++ code with parallel annotations to be no more than 15% slower. These results validate parallelization of the code. All parallel speedup results are presented with respect to the sequential ICC++ code.

5.1 Single Address Space Configuration

The parallel results shown in this section use a Concert virtual machine with all system threads in a single address space. This configuration uses hardware shared memory support for all sharing and communication. Objects can reside in any processors cache, and invalidation of cached data can be issued by any other processor.

Accompanying each speedup graph in this section and Sect. 5.2 will be a corresponding graph of *percent maximum feasible speedup*. The latter graphs show the percentage of maximum feasible speedup that each run of the code achieved. The maximum feasible speedup is less than the ideal speedup because of sequential portions of the SAMR algorithm and because of the limited available parallelism with larger tile sizes. For each parallel execution of the code, the maximum feasible speedup is the sequential execution time divided by the feasible parallel execution time (FPET).

To calculate FPET, we proceeded phase by phase as follows: If a phase is entirely sequential, add its sequential execution time to the FPET. For parallel phases, we considered that the number of grains of available parallelism (e.g. tiles) in a given phase might be less than the number of processors. If a parallel phase has more tiles than processors in the run, then we divide the sequential execution time of the phase by the number of processors and add the result

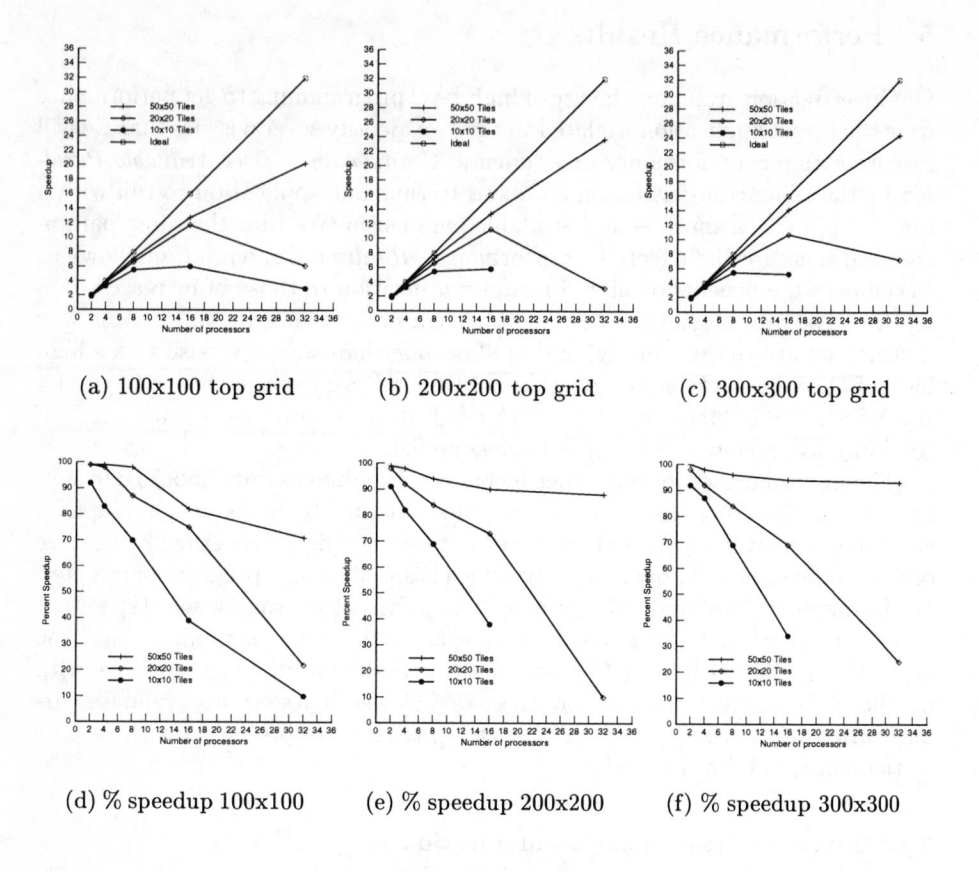

(a) 100x100 top grid   (b) 200x200 top grid   (c) 300x300 top grid

(d) % speedup 100x100   (e) % speedup 200x200   (f) % speedup 300x300

Fig. 2. Multi-level test case

to FPET. Otherwise, we divide the sequential execution time by the maximum number of tiles involved in the phase and add the result to FPET. We emphasize that our calculations use the *maximum* number of tiles operated on by a phase over the entire run as the number of tiles for that phase. Maximum feasible speedup is therefore the maximum speedup that can be achieved by each run of the code, accounting for both sequential portions of the code and lack of parallelism. It is more accurate than ideal speedup as a comparison point to evaluate of Concert parallelization performance.

Figure 2 shows our results for the single address space configuration. The test case run was the adaptive ShockTube 2D test case with a two-level hierarchy. The caption for each figure lists the size of the top-level grid that was run, which approximately determines test case size.

Clearly, tile size affects performance as the code runs on more than four processors. For the 300x300 test case, we see that both 10x10 and 20x20 tile sizes have fallen to less than 70% of feasible speedup on 16 processors. The cost of synchronization and data movement between tiles is limiting performance. The 50x50 tile size cases remain steady achieving 93% of feasible speedup on 32 processors. This last result tells us that we have ample parallelism for up to 32 processors, even for the largest tile size shown here.

Our experiments show that speedup of the tile-to-tile communication phases are the poorest, and tend to limit overall parallel efficiency. See [12] for more details. We do achieve high overall performance, with a high percentage of feasible speedup. This tells us that the shared memory support has aided in parallelization and that the runtime system's load balancing strategy works well with this dynamically parallel code.

5.2 Single Processor Per Address Space Configuration

For the results of this section, the virtual machine configuration creates one system thread per address space. This configuration relies heavily on the runtime messaging layer. For these experiments, we have chosen to run only single-grid, non-adaptive test cases.

Our high-level programming interface allows us to program communication between meshes or tiles as memory copies in our source code, without having to manipulate message passing. For mesh-to-mesh communication, each individual value copied begets a message send in the generated code. This results in a large amount of overhead and synchronization cost between address spaces. In order to limit and measure the magnitude of this problem, we use a technique called *message aggregation*. For the inner copy loop of our communication phase, we simply divide the number of message sends by a constant factor. We perform the aggregation through a source code transformation, which increases sequential time of the ICC++ code by 20% and we assume a constant aggregation factor. The aggregation factor for the experiments shown here is nine, and speedups are shown with respect to the ICC++ code with aggregation, running on one processor.

Figure 3 shows the results of our experiments. We can see that tile size still plays a pivotal role; smaller tiles imply less performance. Compare the speedup of the 50x50 tile with aggregation to that of the 50x50 tile size without. For the 1000x1000 size grid, message aggregation produces good speedups, almost doubling the speedups without aggregation. Note that both the speedup and the actual running time of the transformed code is better than the original on more than eight processors. These results show the benefits of using the high performance messaging with a simple transformation. We achieve close to 90% of feasible speedup with 16 processors for the 1000x1000 case.

5.3 Summary of Results

In this section we saw that we achieved the best absolute performance with the pure shared memory approach and a moderate tile size. This is due mainly to the advanced hardware support for remote memory access being faster than even our low-overhead messaging implementation.

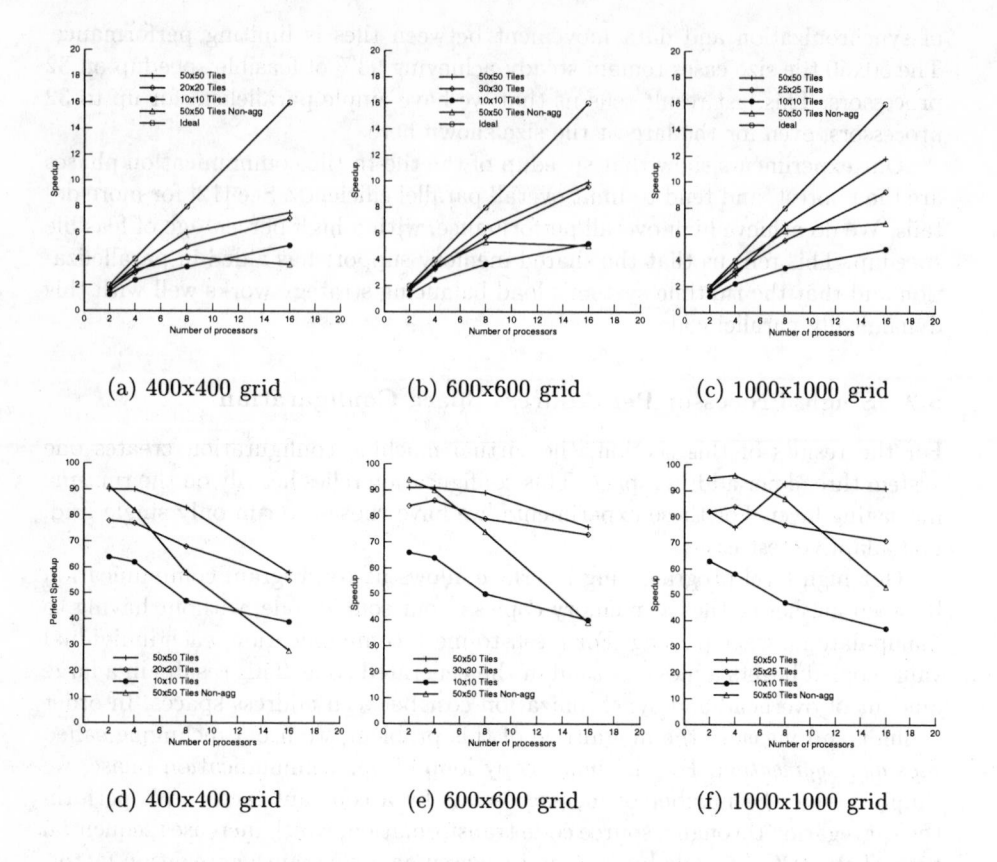

(a) 400x400 grid   (b) 600x600 grid   (c) 1000x1000 grid

(d) 400x400 grid   (e) 600x600 grid   (f) 1000x1000 grid

Fig. 3. Results for single address space per processor, single level hierarchy

The multiple address space runs showed us that a simple, constant factor source code transformation can make use of high performance messaging competitive with expensive shared memory support. The performance improvement with an aggregation factor of nine indicates that an order of magnitude of overhead and synchronization reduction may enable us to use messaging exclusively for communication. This factor would presumably be higher if less efficient messaging primitives were used.

6 Related Work

The Overture project [13] at Los Alamos National Laboratory is also a framework for writing parallel SAMR methods. It is built atop the A++/P++ [14] parallel array class library. Overture is a library of C++ classes designed to represent grids and grid functions. These classes can be instantiated to form a SAMR implementation with a rich set of operations for dealing with different hierarchy topologies and boundary conditions. As Overture is a domain-specific framework, it is restricted to grid-based simulations.

The LPARX [15] parallel system is designed to address the problem of block irregular mesh methods and the problems they present. The system is designed to parallelize methods such as SAMR efficiently, using a collaboration of user and runtime support. LPARX is restricted to regular, non-overlapping grids, and as such is more domain-specific than Overture, and does not approach the richness of data types of an arbitrary object-based system.

7 Summary and Future Work

We achieve up to 24.4 speedup on 32 Origin 2000 processors using maximum shared memory support for communication, which is 93% of maximum feasible speedup for our application. In addition, we get 11.7 speedup on 16 processors for our application by relying heavily on our high-performance messaging support and simple additional techniques. From these results, we can see that the configurable runtime can be used to customize large-scaled shared memory machines to fit specific application behavior, and can therefore potentially be effective on a large range of programs. These results also show that a modest grain size can obtain good performance when utilizing high-performance primitives.

In the future, we plan to expand our experiments on the hybrid distributed-shared memory model. Our single shared address space results suggest that shared memory, in combination with flexible runtime support yields good performance. Our multi-address space results imply that the approach of creating locality regions of memory and passing messages between them can be a powerful systems technology that will scale shared memory machines towards massive parallelism.

Acknowledgments This work is supported in part by DARPA orders #E313 and #E524 through the US Air Force Rome Laboratory Contracts F30602-96-1-0286 and F30602-97-2-0121, and NSF Young Investigator award CCR-94-57809. Support from Microsoft, Intel Corporation, Hewlett-Packard, and Tandem Computers is also gratefully acknowledged.

References

[1] Greg L. Bryan. *The Numerical Simulation of X-ray Clusters*. PhD thesis, University of Illinois at Urbana-Champaign, August 1996.

[2] Andrew Chien, Julian Dolby, Bishwaroop Ganguly, Vijay Karamcheti, and Xingbin Zhang. Supporting high level programming with high performance: The Illinois Concert system. In *Proceedings of the Second International Workshop on High-level Parallel Programming Models and Supportive Environments*, pages 15–24, April 1997.

[3] John Plevyak, Xingbin Zhang, and Andrew A. Chien. Obtaining sequential efficiency in concurrent object-oriented programs. In *Proceedings of the ACM Symposium on the Principles of Programming Languages*, pages 311–321, January 1995.

[4] John Plevyak. *Optimization of Object-Oriented and Concurrent Programs*. PhD thesis, University of Illinois at Urbana-Champaign, Urbana, Illinois, 1996.

[5] Silicon Graphics, Inc., Mountain View, CA. *Origin Servers: Technical Overview of the Origin Family*, 1996. http://www.sgi.com/Products/hardware/servers/technology/overview.html.

[6] Julian Dolby. Automatic inline allocation of objects. In *Proceedings of the 1997 ACM SIGPLAN Conference on Programming Language Design and Implementation*, pages 7–17, Las Vegas, Nevada, June 1997.

[7] Xingbin Zhang and Andrew A. Chien. Dynamic pointer alignment: Tiling and communication optimizations for parallel pointer-based computations. In *Proceedings of ACM SIGPLAN Symposium on Principles and Practice of Parallel Programming*, pages 37–47, Las Vegas, Nevada, June 1997.

[8] Vijay Karamcheti, John Plevyak, and Andrew A. Chien. Runtime mechanisms for efficient dynamic multithreading. *Journal of Parallel and Distributed Computing*, 37(1):21–40, 1996. Available from http://www-csag.cs.uiuc.edu/papers/rtperf.ps.

[9] Gul Agha. *Actors: A Model of Concurrent Computation in Distributed Systems*. MIT Press, Cambridge, MA, 1986.

[10] Vijay Karamcheti. *Run-Time Techniques for Dynamic Multithreaded Computations*. PhD thesis, University of Illinois at Urbana-Champaign, Urbana, Illinois, 1998.

[11] High Performance Fortran Forum. High performance Fortran language specification version 1.0. Technical Report CRPC-TR92225, Rice University, January 1993.

[12] Bishwaroop Ganguly. Concurrent object-oriented programming on large-scaled shared memory: A structured adaptive mesh refinement method in icc++. Master's thesis, The University of Illinois at Urbana-Champaign, Urbana, Illinois, December 1997.

[13] William D. Henshaw. A Primer for Writing PDE Solvers with Overture. La-ur-96-3894, Los Alamos National Laboratory, Los Alamos, NM, December 1997.

[14] R. Parsons and D. Quinlan. A++/P++ array classes for architecture independent finite difference computations. In *Proceedings of the Second Annual Object-Oriented Numerics Conference*, pages 408–418, Sunriver, Oregon, April 1994.

[15] S.R. Kohn and S.B. Baden. Irregular coarse-grain data parallelism under lparx. *Scientific Programming*, 5(3), Fall 1996.

The Matrix Template Library: A Generic Programming Approach to High Performance Numerical Linear Algebra*

Jeremy G. Siek and Andrew Lumsdaine

University of Notre Dame, Notre Dame, IN, USA

Abstract. We present a unified approach for building high-performance numerical linear algebra routines for large classes of dense and sparse matrices. As with the Standard Template Library [1], we separate algorithms from data structures using generic programming techniques. Such an approach does not hinder high performance; rather, writing portable high-performance codes is enabled because the performance-critical code can be isolated from the algorithms and data structures. We address the performance portability problem for architecture-dependent algorithms such as matrix-matrix multiply. Recently, code generation systems, such as PHiPAC [2] and ATLAS [3], have allowed algorithms to be tuned to particular architectures. Our approach is to use *template metaprograms* [4] to directly express performance-critical, architecture-dependent, sections of code.

1 Introduction

Traditional basic linear algebra routines require combinatorial numbers of versions: four precision types (single and double real, single and double complex), several dense storage types (general, banded, packed), a multitude of sparse storage types (13 in the Sparse BLAS Standard Proposal [5]), as well as row and column orientations for each matrix type. A full implementation might require hundreds of versions of the same routine![1] Further, the performance of codes such as matrix-matrix multiply is highly sensitive to memory hierarchy characteristics, so writing portable high-performance codes is even more difficult. A code generation system on top of C or Fortran has been needed to get the flexibility needed for register blocking according to computer architecture.

In this paper we apply fundamental generic programming approaches used by the Standard Template Library (STL) to the domain of numerical linear algebra. The resulting library, the *Matrix Template Library* (MTL), provides comprehensive functionality with a few fundamental algorithms, while also achieving high performance. We explore the use of *template metaprograms* in the construction

---

* This work was supported by NSF grants ASC94-22380 and CCR95-02710.
[1] It is no wonder the NIST implementation of the Sparse BLAS contains over 10,000 routines and an automatic code generation system [6].

```
template <class Row2DIter, class IterX, class IterY> void
matvec::mult(Row2DIter i, Row2DIter iend, IterX x, IterY y) {
  typename Row2DIter::value_type::const_iterator j;
  while (not_at(i, iend)) {
    j = (*i).begin();
    typename IterY::value_type tmp(0);
    while (not_at(j, (*i).end())) {
      tmp += *j * x[j.index()];
      ++j;
    }
    y[i.index()] = tmp;
    ++i;
  }
}
```

Fig. 1. Simplified example of a generic matrix-vector product

of the BLAIS kernels, which provide an elegant solution to portable high performance for matrix-matrix multiply and other blocked codes.

The MTL is in its second generation,[2] the first having been described previously [7]. The current version uses generic programming to a much larger degree than its predecessor.

2 Generic Programming

The principal idea behind the STL is that many algorithms can be abstracted away from the particular data structures on which they operate. Algorithms typically need the abstract functionality of being able to *traverse* through a data structure and *access* its elements. If data structures provide a standard interface for traversal and access, generic algorithms can be mixed and matched with data structures (called *containers* in STL). This interface is realized through the *iterator* (sometimes called a generalized pointer).

Abstractly, linear algebra operations also consist of traversing through vectors and matrices. Vector operations fit neatly into the generic programming approach. The STL already defines several generic algorithms for vectors, such as inner_product(). Extending these generic algorithms to encompass the rest of the Level-1 BLAS [8] is a trivial matter.

Matrix operations are slightly more complex, since the elements are arranged in a 2-dimensional format. The MTL processes matrices as if they are *containers of containers* (note that the matrix implementations are typically not actual containers of containers). The matrix algorithms are coded in terms of *iterators* and *two-dimensional iterators*. A Row2DIter can traverse the rows of a matrix, and produces a row vector when dereferenced. The iterator for the row vector can then be used to access the individual matrix elements. The example in Fig. 1 shows how one can write a generic matrix-vector product.

---

[2] The MTL is available at http://www.lsc.nd.edu/research/mtl/.

Table 1. MTL linear algebra operations

Function Name	Operation	Function Name	Operation
Vector Algorithms		Vector Vector	
set(x,alpha)	$x_i \leftarrow \alpha$	copy(x,y)	$y \leftarrow x$
scale(x,alpha)	$x \leftarrow \alpha x$	swap(x,y)	$y \leftrightarrow x$
s = sum(x)	$s \leftarrow \sum_i x_i$	ele_mult(x,y,z)	$z \leftarrow y \otimes x$
s = one_norm(x)	$s \leftarrow \sum_i \mid x_i \mid$	ele_div(x,y,z)	$z \leftarrow y \oslash x$
s = two_norm(x)	$s \leftarrow (\sum_i x_i^2)^{\frac{1}{2}}$	add(x,y)	$y \leftarrow x + y$
s = inf_norm(x)	$s \leftarrow \max \mid x_i \mid$	s = dot(x,y)	$s \leftarrow x^T \cdot y$
i = find_max_abs(x)	$i \leftarrow$ index of max $\mid x_i \mid$	s = dot_conj(x,y)	$s \leftarrow x^T \cdot \bar{y}$
s = max(x)	$s \leftarrow \max(x_i)$		
s = min(x)	$s \leftarrow \min(x_i)$		
Matrix Algorithms		Matrix Vector	
set(A, alpha)	$A \leftarrow \alpha$	mult(A,x,y)	$y \leftarrow A \times x$
scale(A,alpha)	$A \leftarrow \alpha A$	mult(A,x,y,z)	$z \leftarrow A \times x + y$
set_diag(A,alpha)	$A_{ii} \leftarrow \alpha$	tri_solve(T,x,y)	$y \leftarrow T^{-1} \times x$
s = one_norm(A)	$s \leftarrow max_i(\sum_j \mid a_{ij} \mid)$	rank_one(x,A)	$A \leftarrow x \times y^T + A$
s = inf_norm(A)	$s \leftarrow max_j(\sum_i \mid a_{ij} \mid)$	rank_two(x,y,A)	$A \leftarrow x \times y^T +$
transpose(A)	$A \leftarrow A^T$		$y \times x^T + A$
Matrix Matrix			
copy(A,B)	$B \leftarrow A$	swap(A,B)	$B \leftrightarrow A$
add(A,C)	$C \leftarrow A + C$	ele_mult(A,B,C)	$C \leftarrow B \otimes A$
mult(A,B,C)	$C \leftarrow A \times B$	mult(A,B,C,E)	$E \leftarrow A \times B + C$
tri_solve(T,B,C)	$C \leftarrow T^{-1} \times B$		

3 MTL Algorithms

Table 1 lists the principal algorithms covered by the MTL. This list seems sparse, but a large number of functions are indeed provided through the combination of the above algorithms with the strided(), scaled(), and trans() adapter functions. Figure 2 shows how this is done with a matrix-vector multiply and with a scaled vector assignment.

The unique feature of the MTL is that, for the most part, each of the algorithms is implemented with just one template function. Just one algorithm is used whether the matrix is sparse, dense, banded, single precision, double, complex, etc. From a software maintenance standpoint, the reuse of code gives the MTL a significant advantage over the BLAS [9, 10] or even other object-oriented libraries like TNT [11] (which has different algorithms for different matrix formats).

The generic algorithm code reuse results in the MTL having 10 times fewer lines of code than the Netlib Fortran BLAS while providing greater functionality and achieving generally better performance, especially for level 2 and 3 operations. The MTL has 8,284 lines of code for the algorithms and 6,900 lines of code for dense containers, for a total of 15,184 lines of code. The Fortran BLAS total 154,495 lines of code, an order of magnitude more.

```
//   y <- A * alpha x
matvec::mult(trans(A), scaled(x, alpha), strided(y,incy));
//   y <- alpha x
vecvec::copy(scaled(x, alpha), y);
```

Fig. 2. Transpose, scaled, and strided adapters

4   MTL Components

The MTL defines a set of data structures and other components for representing linear algebra objects. An MTL matrix is constructed with layers of components. Each layer is a collection of classes that are templated on the lower layer. The bottom most layer consists of the numerical types (float, double, etc). The next layers consist of 1-D containers followed by 2-D containers. The 2-D containers are wrapped up with an *orientation*, which in turn is wrapped with a *shape*. A complete MTL matrix type typically consists of a templated expression in the form

$$shape<orientation<twod<oned<num\_type> > > >$$

For example, an upper triangular matrix would be defined as

$$triangle < column < dense2D < double >>, upper >$$

Some 2-D containers also subsume the 1-D type, such as the contiguous dense2D container.

Matrix Orientation The row and column adapters map the *major* and *minor* aspects of a matrix to the corresponding *row* or *column*. This technique allows the same code for data structures to provide both row and column orientations of the matrix. 2-D containers must be wrapped up with one of these adapters to be used in the MTL algorithms.

Matrix Shape Matrices can be categorized into several shapes: general, upper triangular, lower triangular, symmetric, Hermitian, etc. The traditional approach to handling the algorithmic differences due to shape is to have a separate function for each type. For instance, in the BLAS we have a _GEMV, _SYMV, _TRMV, etc. The MTL instead uses different data structures for each shape, with the banded, triangle, symmetric, and hermitian matrix adapters. It is the responsibility of these adapters to make sure that they work with all of the MTL generic algorithms. The MTL philosophy is to use *smarter* data structures to allow for fewer and simpler algorithms.

5   The High Performance Layer

We have presented many levels of abstraction, and a set of unified algorithms for a variety of matrices, but high performance must be achieved. Template-based programming coupled with modern compilers such as KAI C++ [12] provide several mechanisms for high-performance.

Static Polymorphism The template facilities in C++ allow functions to be selected at compile-time based on data type. This provides a mechanism for abstraction which preserves high performance. Dynamic (run-time) dispatch is avoided, and the template functions can be inlined just as regular functions. This ensures that the numerous small function calls in the MTL (such as iterator increment operators) introduce no extra overhead.

Lightweight Object Optimization The generic programming style introduces a large number of small objects into the code. This incurs a performance penalty because the presence of a structure can interfere with other optimizations, including the mapping of the individual data items to registers. This problem is solved with small object optimization, also know as scalar replacement of aggregates [13], which is performed by the KAI C++ compiler.

Automatic Unrolling Modern compilers do a great job of unrolling loops and scheduling instructions, but typically only for recognizable cases. There are many ways, especially in C and C++, to interfere with the optimization process. The abstractions of the MTL are designed to result in code that is easy for the compiler to optimize. Furthermore, the *iterator* abstraction makes inter-compiler portability possible, since it encapsulates how looping is performed.

Algorithmic Blocking The bane of portable high performance numerical linear algebra is the need to tailor key routines to specific execution environments. For example, to obtain high performance on a modern microprocessor, an algorithm must properly exploit the memory hierarchy and pipeline architecture (typically through careful loop blocking and structuring). Ideally, one would like to express high performance algorithms in a portable fashion, but there is not enough expressiveness in languages such as C or Fortran to do so. Recent efforts (PHiPAC [2], ATLAS [3]) have resorted to going outside the language, i.e., to code generation systems, in order to gain this kind of flexibility. The Basic Linear Algebra Instruction Set (BLAIS) is a library specification that takes advantage of C++ features to express high-performance loop structures at a high level.

5.1 The Basic Linear Algebra Instruction Set (BLAIS)

The BLAIS specification contains *fixed-size* algorithms with functionality equivalent to that of the Level-1, Level-2, and Level-3 BLAS [9, 10, 8]. The BLAIS routines themselves are implemented using the Fixed Algorithm Size Template (FAST) library, which contains general purpose fixed-size algorithms equivalent in functionality to the generic algorithms in the STL. The thin BLAIS routines map the generic FAST algorithms into fixed-size mathematical operations. There is no added overhead in the layering because all the function calls are inlined. Using the FAST library allows the BLAIS routines to be expressed in a simple and elegant fashion. Note that the intended use of the BLAIS routines is to carry out the register blocking within a larger algorithm. This means the BLAIS routines handle only small matrices, and therefore avoid the problem of excessive code bloat.

```
int x[4] = {1,1,1,1}, y[4] = {2,2,2,2};

// STL
template <class InIter1, InIter2, OutIter, BinaryOp>
OutIter transform(InIter1 first1,InIter1 last1,InIter2 first2,
                  OutIter result,BinaryOp binary_op);

transform(x, x + 4, y, y, plus<int>());

// FAST
template <int N, class InIter1, class InIter2,
          class OutIter, class BinOp>
OutIter fast::transform(InIter1 first1, cnt<N>,InIter2 first2,
                        OutIter result, BinOp binary_op);

fast::transform(x, cnt<4>(), y, y, plus<int>());
```

Fig. 3. Example usage of STL and FAST versions of `transform()`

We describe the FAST algorithms and show how the BLAIS are constructed from them. We show how the BLAIS can be used as high-level instructions (kernels) to handle the register-level blocking in a matrix-matrix product. Experimental results show that the performance obtained by our approach can equal and even exceed that of vendor-tuned libraries.

```
// The general case
template <int N, class InIter1, class InIter2,
          class OutIter, class BinOp>
inline OutIter
fast::transform (InIter1 first1, cnt<N>, InIter2 first2,
                 OutIter result, BinOp binary_op) {
  *result = binary_op (*first1, *first2);
  return transform(++first1, cnt<N-1>(), ++first2,
                   ++result, binary_op);
}
// The N = 0 case to stop template recursion
template<class InItr1,class InItr2,class OutItr,class BinOp>
inline OutItr
fast::transform (InItr1 first1, cnt<0>, InItr2 first2,
                 OutItr result, BinOp binary_op) {
  return result; }
```

Fig. 4. Definition of FAST `transform()`

Fixed Algorithm Size Template (FAST) Library The FAST Library includes generic algorithms such as `transform()`, `for_each()`, `inner_product()`, and accumulate() that are found in the STL. The interface closely follows that of the STL. All input is in the form of *iterators*. The only difference is that the loop-end iterator is replaced by a *count template* object. The example shown in Fig. 3 demonstrates the use of both the STL and FAST versions of transform() to realize an AXPY-like operation $(y \leftarrow x + y)$. The first1 and last1 parameters are iterators for the first input container (indicating the beginning and end of the container, respectively). The first2 parameter is an iterator indicating the beginning of the second input container. The result parameter is an iterator indicating the start of the output container. The binary_op parameter is a function object that combines the elements from the first and second input containers into the result containers.

```
// Definition
template <int N> struct vecvec::add {
  template <class Iter1, class Iter2> inline
  vecvec::add(Iter1 x, Iter2 y) {
    typedef typename iterator_traits<Iter1>::value_type T;
    fast::transform(x, cnt<N>(), y, y, plus<T>());
}};
// Example use
double x[4], y[4];                    // y[0] += a * x[0];
fill(x, x+4, 1); fill(y, y+4, 5);     // y[1] += a * x[1];
double a = 2;                         // y[2] += a * x[2];
vecvec::add<4>(scl(x, a), y);         // y[3] += a * x[3];
```

Fig. 5. Definition and use of BLAIS add()

```
// General Case
template <int M, int N>
struct mult {
  template <class AColIter, class IterX, class IterY> inline
  mult(AColIter A_2Diter, IterX x, IterY y) {
    vecvec::add<M>(scl((*A_2Diter).begin(), *x), y);
    mult<M, N-1>(++A_2Diter, ++x, y);
  }
};
// N = 0 Case
template <int M>
struct mult<M, 0> {
  template <class AColIter, class IterX, class IterY> inline
  mult(AColIter A_2Diter, IterX x, IterY y) {
    // do nothing
  }
};
```

Fig. 6. BLAIS matrix-vector multiplication

The difference between the STL and FAST algorithms is that STL accommodates containers of arbitrary size, with the size being specified at run-time. FAST also works with containers of arbitrary size, but the size is fixed at compile time. In Fig. 4, we show how the FAST `transform()` routine is implemented. We use a tail-recursive algorithm to achieve complete unrolling — there is no actual loop in the FAST `transform()`. The template-recursive calls are inlined, resulting in a sequence of N copies of the inner loop statement. This technique (sometimes called *template metaprograms*) has been used to a large degree in the Blitz++ Library [14].

BLAIS Vector-Vector Operations Figure 5 gives the implementation for the BLAIS vector `add()` routine, and shows an example of its use. The FAST `transform()` algorithm is used to carry out the vector-vector addition as it was in the example above.

The comments on the right show the resulting code after the call to `add()` is inlined. The `scl()` function used above demonstrates the purpose of the `scale_iterator`. The `scale_iterator` multiplies the value from x by a when the iterator is dereferenced within the `add()` routine. This adds no extra time or space overhead due to inlining and lightweight object optimizations. The `scl(x, a)` call automatically creates the proper `scale_iterator` out of x and a.

BLAIS Matrix-Vector Operations The BLAIS matrix-vector multiply implementation is depicted in Fig. 6. The algorithm simply carries out the vector add operation for the columns of the matrix. Again a fixed depth recursize algorithm is used, which becomes inlined by the compiler.

BLAIS Matrix-Matrix Operations The BLAIS matrix-matrix multiply is implemented using the BLAIS matrix-vector operation. The code looks very similar to the matrix vector multiply, except that there are three integer template arguments (M, N, and K), and the inner "loop" contains a call to `matvec::mult()` instead of `vecvec::add()`.

5.2 A Configurable Recursive Matrix-Matrix Multiply

A high performance matrix-matrix multiply code is highly sensitive to the memory hierarchy of a machine, from the number of registers to the levels and sizes of cache. For highest performance, algorithmic blocking must be done at each level of the memory hierarchy. A natural way to formulate this is to write the matrix-matrix multiply in a recursive fashion, where each level of recursion performs blocking for a particular level of the memory hierarchy.

We take this approach in the MTL algorithm. The size and shapes of the blocks at each level are determined by the *blocking adapter*. Each adapter contains the information for the next level of blocking. In this way the recursive algorithm is determined by a recursive template data-structure (set up at compile time). The setup code for the matrix-matrix multiply is shown in Fig. 7. This example blocks for just one level of cache, with 64 x 64 blocks. The small 4 x

```
template <class MatA, class MatB, class MatC>
void matmat::mult(MatA& A, MatB& B, MatC& C) {
  MatA::RegisterBlock<4,1> A_L0;   MatA::Block<64,64> A_L1;
  MatB::RegisterBlock<1,2> B_L0;   MatB::Block<64,64> B_L1;
  MatC::CopyBlock<4,2> C_L0;       MatC::Block<64,64> C_L1;
  matmat::__mult(block(block(A, A_L0), A_L1),
                 block(block(B, B_L0), B_L1),
                 block(block(C, C_L0), C_L1));
}
```

Fig. 7. Setup for the recursive matrix-matrix product

```
template <class MatA, class MatB, class MatC>
void matmat::__mult(MatA& A, MatB& B, MatC& C) {
  A_k = A.begin_columns(); B_k = B.begin_rows();
  while (not_at(A_k, A.end_columns())) {
    C_i = C.begin_rows(); A_ki = (*A_k).begin();
    while (not_at(C_i, C.end_rows())) {
      B_kj = (*B_k).begin(); C_ij = (*C_i).begin();
      MatA::Block A_block = *A_ki;
      while (not_at(B_kj, (*B_k).end())) {
        __mult(A_block, *B_kj, *C_ij);
        ++B_kj; ++C_ij;
      } ++C_i; ++A_ki;
    } ++A_k; ++B_k;
  }
}
```

Fig. 8. A recursive matrix-matrix product algorithm

2 blocks fit into registers. Note that these numbers would normally be constants that are set in a header file.

The recursive algorithm is listed in Fig. 8. The bottom recursion level is implemented with a separate function that uses the BLAIS matrix-matrix multiply, and "cleans up" the leftover edge pieces.

5.3 Optimizing Cache Conflict Misses

Besides blocking, another optimization for matrix-matrix multiply code is block copying. Typically utilization of the level-1 cache is much lower than one might expect due to cache conflict misses. This is especially apparent in direct-mapped and low associativity caches. This problem is minimized copying the current block of matrix A into a contiguous section of memory [15], allowing blocking sizes closer to the size of the L-1 cache without inducing as many cache conflict misses.

It turns out that this optimization is straightforward to implement in our recursive matrix-matrix multiply. We already have block objects (submatrices A_block, *B_j, and *C_j) in Fig. 8. We modify the constructors for these objects

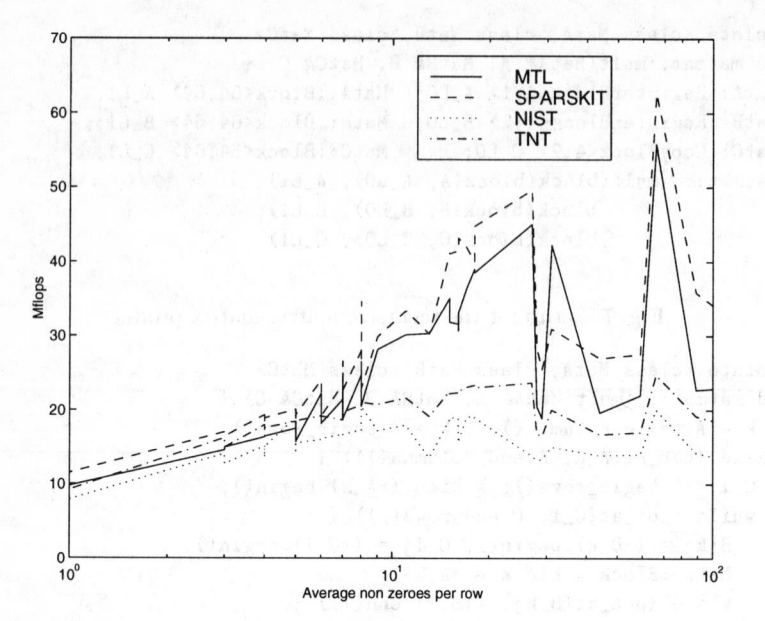

Fig. 9. Performance results for sparse matrix-vector multiply

to make a copy to a contiguous part of memory, and the destructors to copy the block back to the original matrix. This is especially nice since the optimization does not clutter the algorithm code, but instead the change is encapsulated in the copy_block matrix class.

6 Performance Results

We present performance results comparing the MTL to several public-domain and vendor-tuned numerical libraries. Timings were obtained on a Sun Ultra-SPARC 170E workstation using KAI C++ [12] (for C++ to C translation) and the Solaris C compiler with maximum available optimizations. Figure 9 shows performance results for matrix-vector product computation using an assortment of sparse matrices from the MatrixMarket [16]. Results are shown for the MTL, SPARSKIT [17] (Fortran), NIST [6](C), and TNT (C++), using row-major compressed storage. Performance results for dense matrix-matrix multiply are shown in Fig. 10, where we compare the MTL, the Sun Performance Library, TNT, and the Netlib Fortran BLAS, using column-major storage.

7 Supplemental Libraries

The MTL provides an extensive foundation for other portable high-performance libraries. We have created two: the Iterative template Library (ITL), and an implementation of the legacy BLAS. The ITL is a collection of iterative solvers

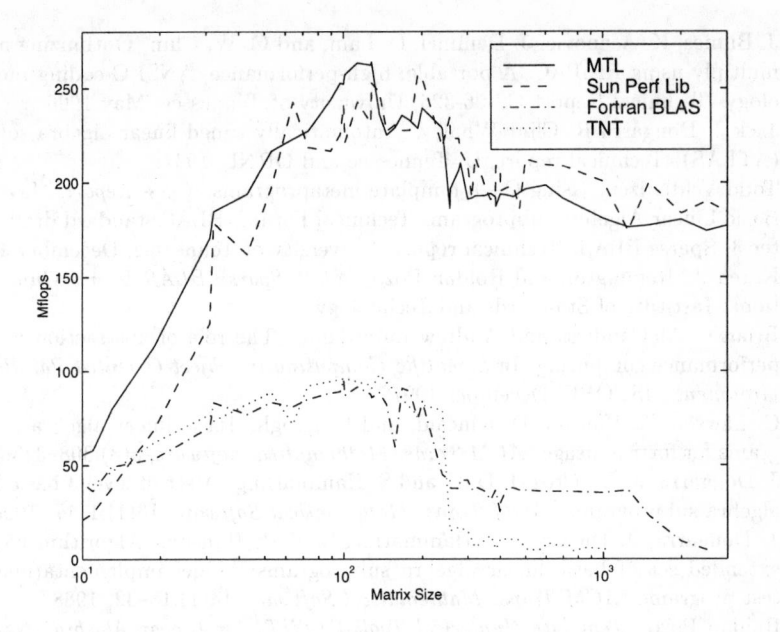

Fig. 10. Performance results for dense matrix-matrix multiply

(similar to the Iterative Methods Library [18]) that uses the MTL for its basic linear algebra operations. Our legacy BLAS implementation is a simple Fortran-callable interface to the MTL data structures and algorithms. We have also provided an MTL interface to LAPACK [19], so that users of the MTL have a convenient way to access the LAPACK functionality.

8 Conclusion

Attempts to create portable high performance linear algebra routines have used specialized code generation scripts to provide enough flexibility in C and Fortran. We have shown that C++ has enough expressiveness to allow codes to be reconfigured for particular architectures by merely changing a few constants. Further, advanced C++ compilers can still aggressively optimize in the presence of the powerful MTL abstractions, producing code that matches or exceeds the performance of hand-coded C and vendor-tuned libraries.

Acknowledgments This work was supported by NSF grants ASC94-22380 and CCR95-02710. The authors would like to express their appreciation to Tony Skjellum and Puri Bangalore for numerous helpful discussions.

References

[1] Meng Lee and Alexander Stepanov. The standard template library. Technical report, HP Laboratories, February 1995.

[2] J. Bilmes, K. Asanovic, J. Demmel, D. Lam, and C.-W. Chin. Optimizing matrix multiply using PHiPAC: A portable, high-performance, ANSI C coding methodology. Technical Report CS-96-326, University of Tennessee, May 1996.

[3] Jack J. Dongarra R. Clint Whaley. Automatically tuned linear algebra software (ATLAS). Technical report, U. Tennessee and ORNL, 1997.

[4] Todd Veldhuizen. Using C++ template metaprograms. *C++ Report*, May 1995.

[5] Basic Linear Algebra Subprograms Technical Forum. BLAS standard draft chapter 3: Sparse BLAS. Technical report, University of Tennessee, December 1997.

[6] Karen A. Remington and Roldan Pozo. *NIST Sparse BLAS User's Guide*. National Institute of Standards and Technology.

[7] Brian C. McCandless and Andrew Lumsdaine. The role of abstraction in high-performance computing. In *Scientific Computing in Object-Oriented Parallel Environments*. ISCOPE, December 1997.

[8] C. Lawson, R. Hanson, D. Kincaid, and F. Krogh. Basic linear algebra subprograms for fortran usage. *ACM Trans. Mathematical Software*, 5(3):308–323, 1979.

[9] J. Dongarra, J. Du Croz, I. Duff, and S. Hammarling. A set of level 3 basic linear algebra subprograms. *ACM Trans. Mathematical Software*, 16(1):1–17, 1990.

[10] J. Dongarra, J. Du Croz, S. Hammarling, and R. Hanson. Algorithm 656: An extended set of basic linear algebra subprograms: Model implementations and test programs. *ACM Trans. Mathematical Software*, 14(1):18–32, 1988.

[11] Roldan Pozo. *Template Numerical Toolkit (TNT) for Linear Algebra*. National Insitute of Standards and Technology.

[12] Kuck and Associates. *Kuck and Associates C++ User's Guide*.

[13] Steven Muchnick. *Advanced Compiler Design and Implementation*. Morgan Kaufmann, 1997.

[14] Todd Veldhuizen and M. Ed Jernigan. Will C++ be faster than Fortran. In *Scientific Computing in Object-Oriented Parallel Environments*. ISCOPE, December 1997.

[15] Monica S. Lam, Edward E. Rothberg, and Michael E. Wolf. The cache performance and optimizations of blocked algorithms. In *ASPLOS IV*, April 1991.

[16] NIST. MatrixMarket. http://gams.nist.gov/MatrixMarket/.

[17] Youcef Saad. SPARSKIT: a basic tool kit for sparse matrix computations. Technical report, NASA Ames Research Center, 1990.

[18] Roldan Pozo Jack Dongarra, Andrew Lumsdaine and Karin A. Remington. *Iterative Methods Library Reference Guide*, v. 1.2 edition, April 1997.

[19] E. Anderson, Z. Bai, C. Bischoff, J. Demmel, J. Dongarra, J. DuCroz, A. Greenbaum, S. Hammarling, A. McKenney, and D. Sorensen. LAPACK: A portable linear algebra package for high-performance computers. In *Proceedings of Supercomputing '90*, pages 1–10. IEEE Press, 1990.

The Mobile Object Layer: A Run-Time Substrate for Mobile Adaptive Computations

Nikos Chrisochoides*, Kevin Barker**, Démian Nave***, and Chris Hawblitzel†

University of Notre Dame, Notre Dame, IN, USA

Abstract. We present a parallel runtime substrate that supports a global addressing scheme, object mobility, and automatic message forwarding required for the implementation of adaptive applications on distributed memory machines. Our approach is application-driven; the target applications are characterized by very large variations in time and length scales. Preliminary performance data from parallel unstructured adaptive mesh refinement on an SP2 suggest that the flexibility and general nature of the approach we follow does not cause undue overhead.

1 Introduction

We present a lean, language-independent, and easy to port and maintain runtime system for the efficient implementation of adaptive applications on large-scale parallel systems. Figure 1 depicts the architecture of the overall system and its layers that address the different requirements of such an application.

The first layer, the Data-Movement and Control Substrate (DMCS) [1] provides thread-safe, one-sided communication. DMCS implements an application programming interface (API) proposed by the PORTS consortium; it resembles Nexus [2] and Tulip [3], two other mid-level communication systems that implement similar API's with different design philosophies and objectives. It has been implemented on IBM's Low-level Application Programming Interface (LAPI) [4] on the SP2. For data-movement operations, our measurements on an SP2 show that the overhead of DMCS is very close (within 10% for both puts and gets) of the overhead of LAPI.

The second layer, the Mobile Object Layer (MOL), will be described in more detail in Sect. 2. The MOL supports a global addressing scheme designed for object mobility, and provides a correct and efficient protocol for message forwarding and communication between migrating objects. We describe a parallel adaptive mesh generator which uses the MOL as the run-time system for dynamic load balancing and message passing in Sect. 3. Preliminary performance data (Sect. 4) suggest that the flexibility and general nature of the MOL's approach for data migration do not cause undue overhead. However, the MOL does not provide

* Supported by NSF grant #9726388 and JPL award #961097
** Supported by the Arthur J. Schmidt Fellowship and by NSF grant #9726388
*** Supported by NSF grant #9726388
† Supported by an NSF fellowship for a portion of this work.

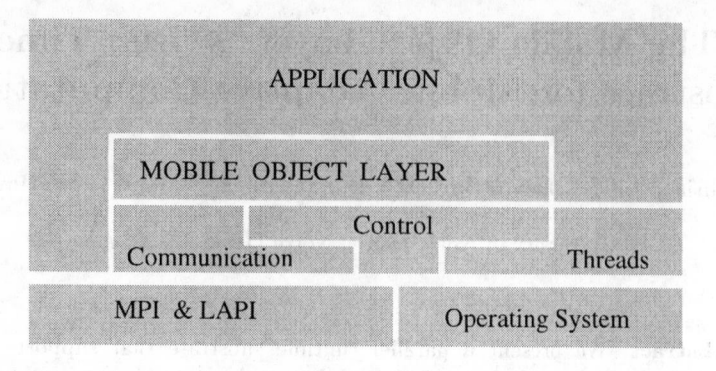

Fig. 1. Parallel run-time system: architecture

support for efficient shared memory management, or implement policies specifying how and when mobile objects must be moved. The MOL implements some functionality provided by distributed-shared memory (DSM) systems (Sect. 5). We conclude with a summary of our current work, and a description of plans to improve the functionality of the MOL, in Sects. 6 and 7.

2   The Mobile Object Layer

The Mobile Object Layer provides tools to build migratable, distributed data structures consisting of *mobile objects* linked via *mobile pointers* to these objects. For example, a distributed graph can be constructed using mobile objects for nodes and mobile pointers to point to neighboring nodes. If a node in such a structure is moved from one processor to another, the MOL guarantees [5] that messages sent to the node will reach it by forwarding them to the node's new location. The MOL's forwarding mechanism assumes no network ordering and allows network delays of arbitrary length to halt message reception. Also, forwarding only affects the source and target processors of a message to a mobile object; this "lazy" updating minimizes the communication cost of moving an object. The sequence number contained in the MoveInfo structure passed with a mobile object is compared to the sequence number contained in the target processors local directory [6]. In this way, the MOL prevents old updates from overwriting newer ones.

The MOL uses a distributed directory structure which allows fast local access to the locations of mobile objects. Each processor maintains its own directory; each directory entry corresponds to the processor's "best guess" of the corresponding object's location. Compared to a central directory, this method reduces network traffic, but introduces the problem of maintaining global directory consistency in the presence of object migration. As an alternative to the expensive, non-scalable method of broadcasting directory updates to all processors, the MOL implements a "lazy" updating scheme, allowing some directory entries to be out of date. A processor communicates with a particular mobile object by sending messages to the "best guess" location given by its local directory. If this location is incorrect, the sending processor is informed of the object's true location; only processors that show an explicit interest in an object are updated with the object's correct location.

Although the MOL provides mechanisms to support mobile objects, there are no policies specifying how, when, and to where the mobile objects must be moved. It is the responsibility of application-specific software to coordinate object migration. This allows the MOL to support many systems, since no single migration policy could efficiently satisfy the needs of a broad range of irregular, parallel applications. The MOL's flexibilty and low-overhead interface make it an efficient run-time system on which application-specific libraries and languages can be built.

2.1 Mobile Pointers and Distributed Directories

The basic building block provided by the MOL is the "mobile pointer." A mobile pointer consists of two integer numbers: a 16-bit processor number, which specifies where the corresponding object was originally allocated (the "home node"), and a 32-bit index number which is unique on the object's home node. This pair forms an identifier for a mobile object which is unique on every processor in the system and which can be passed as data in messages without extra help from the MOL. Using the mobile pointer, the corresponding object's "best guess" location can be retrieved from a processor's directory.

A directory is a two dimensional array of directory entries; an entry for any mobile pointer can be located by indexing into the directory with the mobile pointer's home node and index number. A mobile object's directory entry consists of a 16-bit processor number containing the object's "best guess" location, a 16-bit sequence number indicating how up to date the best guess is, and a 32-bit physical pointer to the object's data. The pointer can be retrieved using *mob_deref()*, which returns NULL if the object is not physically located on the requesting processor.

There are three possibilities for the state of an object's location as shown by the object's directory entry. First, the object may reside on the current processor, in which case the message can be handled locally. Second, the object may reside at a remote processor; in this case, the message is sent to the processor indicated by the directory entry. If the target processor does not contain the object, it will forward the message to the best guess location given by the object's local directory entry. Third, the directory may not have an entry for the mobile pointer. In this case, the mobile pointer's home node is used as the default best guess location for the object.

2.2 Message Layer

Our implementation of the MOL builds its own message layer on top of the Data Movement and Control Substrate (DMCS), which is in turn built on top of IBM's Low-level API (LAPI) [4]. In order to make the MOL as flexible and as portable as possible, versions also exist which use Active Messages (AM) [5] or NEXUS [2] as the underlying transport mechanism. In our implementation, we use incoming and outgoing pools of messages, similar to those used in the Generic Active Message specification from Berkeley [7]. The MOL supports "processor requests" with *mob_request()* and "object messages" with *mob_message()*, to transfer messages smaller than 1024 bytes to processors and mobile objects, respectively. 1024 bytes was empirically analytically found to be the maximum message size for which store-and-forwarding is more efficient than three-way-rendezvous for forwarded messages on the SP2. Both types require a user-supplied procedure to handle the message when it arrives at its destination.

Three types of handlers are available to process a message initiated by the MOL. The first type is a function handler, which is similar to an AM handler. This is the fastest handler type, since it is executed immediately upon being received and processed by the MOL, but neither communication nor context switching is allowed within the handler. Second, the message may be processed from within a delayed handler, which is queued internally by the MOL. The delayed handler is slower but also more flexible, in that communication, but not context switching, is allowed from inside the handler. Third, a threaded handler can spawn a thread to process the message. This is the most flexible handler type, but it is also the slowest. Since each of these handlers may be appropriate in different situations, the MOL supports all three; the type of handler is specified as an argument to *mob_request()* and *mob_message()*. All handlers are passed the sending processor number, the physical address and length of the message data, and one user-defined argument. In addition, object message handlers are passed the mobile pointer and the local address of the mobile object.

Although the MOL only directly supports small and medium sized messages (store and forwarding of large messages is inefficient) efficient large message protocols can be built using MOL messaging. As a simple example, suppose the user wishes to create a large "get" operation directed to a mobile object. This can be done simply by creating a local buffer to hold incoming data, and then sending an object message including the buffer's address to the target object. The delayed or threaded remote handler can then call a store procedure like AM's *am_store()* to save the requested data to the buffer in the originating process.

Large message send/receive protocols, typically accomplished with a three way rendezvous, can also be implemented with the MOL. As in the first example, an object message is sent to the target object containing the amount of space to allocate. The remote handler then allocates the buffer and sends a request to the originating processor with the buffer's address. Finally, the source processor transfers the data, using a store procedure such as *am_store()*. In this example, the object must be "locked" by the programmer to keep the object from moving before the store operation completes, since the MOL does not control the migration of objects.

3 Application: Parallel Grid Generation

The efficient implementation of an unstructured mesh generator on distributed memory multiprocessors requires the maintenance of complex and dynamic distributed data structures for tolerating latency, minimizing communication and balancing processor workload. We describe an implementation of the parallel 2D Constrained Delaunay Triangulation (CDT) which uses the Mobile Object Layer to simplify data migration for load balancing the computation. We have chosen a simple work-stealing method [8, 9] to demonstrate the effectiveness of the MOL as the bookkeeper for message-passing when data is migrated by the load balancing module.

Constrained Delaunay Triangulation. The mesh generator uses a Constrained Delaunay Triangulation method [10] to generate a guaranteed-quality mesh [11]. Given a precomputed domain decomposition, each subdomain is refined independently of the other regions, except at the interfaces between regions. For 2D meshes, the extent of the refinement is defined by "constrained" interface and boundary edges. If a boundary or interface edge is part of a triangle to be refined, that edge is split. Since interface edges are shared between regions, splitting an edge in one region causes the change to propagate to the region which shares the split edge. The target region is updated as if it had split the edge itself.

Load Balancing with the MOL. The input to the mesh generator is a decomposition of a domain into some number of regions, which are assigned to processors in a way that maximizes data locality. Each processor is responsible for managing multiple regions, since, in general, there will be an over-decomposition of the domain. Subsequently, imbalance can arise due to both unequal distribution of regions and large differences in computation (e.g. between high- and low-accuracy regions in the solution).

The work-stealing load balancing method we implement maintains a counter of the number of work-units that are currently waiting to be processed, and consults a threshhold of work to determine when work should be requested from other processors [12]. When the number of work-units falls below the threshhold, a processor requests a sufficient amount of work to maintain consistent resource utlization.

The regions can be viewed as the work-units or objects which the load balancer can migrate to rebalance the computation. Using the MOL, each region is viewed as a mobile object; we associate a mobile pointer with each region, which allows messages sent to migrated regions to be forwarded to the new locations. The load balancing module can therefore migrate data without disrupting the message-passing in the computation.

Data Movement Using the MOL. The critical steps in the load balancing phase are the region migration, and the updates for the edge split messages between regions. To move a region, the MOL requires that *mob_uninstallObj* be called to update the sending processor's local directory to reflect the pending change in the region's location. Next, a programmer-supplied procedure is used to pack the region's data into a buffer, which must also contain the region's mobile pointer and the 4-byte MoveInfo structure returned by *mob_uninstallObj* to track the region's migration. Then, a message-passing primitive (e.g. *MPI_SEND*) is invoked to transport the buffer, and another user-supplied procedure unpacks and rebuilds the region in the new processor. After the region has been unpacked, *mob_installObj* must be called with the region's mobile pointer and the MoveInfo structure to update the new processor's directory.

Since the MOL is used to move data, standard message-passing primitives, like *MPI_SEND*, will not work to send split edge messages from one region to another, since regions can be migrated. The MOL will forward a split edge message sent with *mob_message*, and will update the sending processor's directory so that, unless the target region moves again, subsequent messages will not be forwarded.

4   Preliminary Performance Data

We present results for *mob_message*, which allows messages to be sent to a mobile object via a mobile pointer, and for *mob_request*, which directs messages of 1024 bytes or less (a parameterized value) to specific processors without explicitly requesting storage space on the target processor. In addition, we present data gathered from the parallel meshing application for both non-load balanced and load balanced runs at different percentages of imbalance in the computation.

All measurements for *mob_message* and *mob_request* were taken on an IBM RISC System/6000 SP, using Active Messages [13]. The benchmarks measured the per-hop latency of messages ranging from 8 to 1024 bytes, as compared to the equivalent *am_store* calls. The performance is very reasonable; the latency of *mob_request* is within about 11% of the latency of *am_store*, while *mob_message*'s latency is about 12% to 14% higher than *am_store*'s latency.

To illustrate the importance of the MOL's updates, Fig. 2 shows the latency of messages that were forwarded once each time they were sent versus messages that were not forwarded. Not surprisingly, the latency of the forwarded messages was about twice as high as that of the unforwarded messages. In a real application, the overall (amortized) cost of forwarding is determined by how often an object moves versus how often messages are sent to the object, since messages are forwarded immediately after an object moves but not after the updates have been received. In the case of the mesh generator, a large number of split edge requests are sent, relative to the number of times a mesh region is migrated (see Fig. 5), resulting in a low amortized cost for forwarding split-edge requests. Figure 3 shows the performance of the MOL's three types of handlers. The graph clearly shows that the overheads caused by the delayed and threaded handlers are fairly low relative to the functionality they add.

The next set of graphs represents data for a parallel mesh with between 100,000 and 170,000 elements, and for load imbalances of between 8 and 50 percent. Each of the four processors in the system started with 16 regions. All

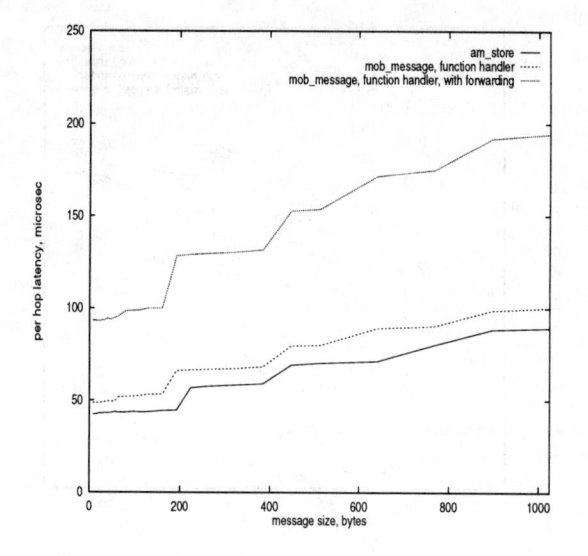

Fig. 2. Mobile object layer performance: forwarding overhead

measurements were taken on an SP2, using a NEXUS implementation of the MOL.

Figure 4 represents the minimum and maximum computation times on the four processors in the non-load balanced experiments. Given above each bar is the number of elements generated in the mesh for that particular run. Figure 5 displays the maximum computation time for a series of load-balanced mesh computations which used the same initial mesh as the non-load balanced experiments. Each bar is broken down into the time spent triangulating regions, packing and servicing split edge requests, and forwarding messages to migrated regions, in order to show the minimal overhead of using the MOL's forwarding mechanism. The tuple above each bar gives the number of split edge requests and the number of object migrations.

5   Related Work

Our run-time system provides global address space and supports object mobility, as do many other previously developed systems and high performance languages designed for irregular applications. Examples of such systems and languages are: ABC++ [14], TreadMarks [15], Charm++ [16], Chaos++ [17], CC++ [18], Amber [19] and CRL [20] to mention a few.

Chaos++ [17] supports *globally addressable objects* an abstraction similar to *mobile objects*. In Chaos++ global objects are owned by a single processor and all other processors with data-dependencies to a global object possess shadow copies of the global object. The Mobile Object Layer does not use shadow objects; instead, it relies on an efficient message-forwarding mechanism to locate and fetch data from remote objects.

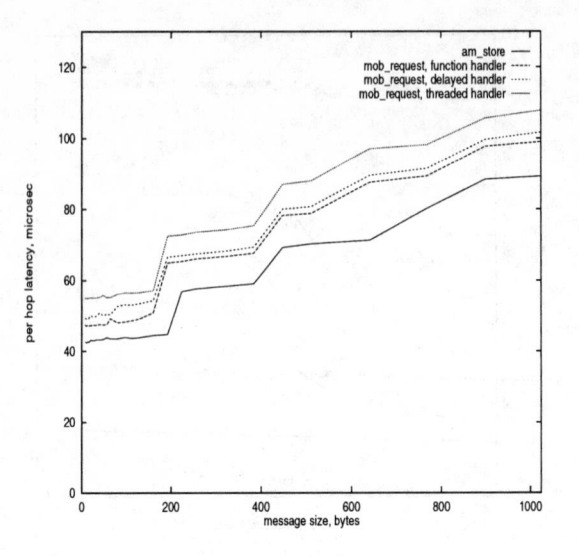

Fig. 3. Mobile object layer performance: handler overhead

ABC++ [14] proposed an object migration mechanism that would allow an object to move away from its original "home node." However, the proposed mechanism would have required communication with the home node each time a message is sent to the object. The MOL eliminates additional communication for every message, because its directories are automatically updated to keep track of where objects have migrated. Furthermore, MOL updates are not broadcast to all processors in the system, but are lazily sent out to individual processors as needed. The MOL protocol for dealing with updates correctly and efficiently is nontrivial, and goes beyond the proposals presented for ABC++.

FLASH [21] integrates both messge passing and shared memory into a single architecture. The key feature of the FLASH architecture is the MAGIC programmable node controller which connects processor, memory, and network components at each node. MAGIC is an embedded processor which can be programmed to implement both cache coherence and message passing protocols. The Mobile Object Layer, on the other hand, is designed to be a thin software layer which can exist independently from the underlying hardware. The MOL is isolated from the system hardware by the DMCS layer, which allows the MOL to be portable as well as tunable; the MOL can be tuned to extract maximal performance by providing a vendor-specific DMCS implementation.

The MOL offers a substantial improvement over explicit message passing systems, such as systems built using MPI. The primary functionality added by the MOL is the ability to send messages to *mobile objects*, not processors. In other words, with the MOL, it is possible to communicate with objects without knowledge of the object's location. This greatly eases the burden placed on the developers of mobile, adaptive applications. The MOL hides the complexity involved with maintaining the validity of global pointers by employing a sepa-

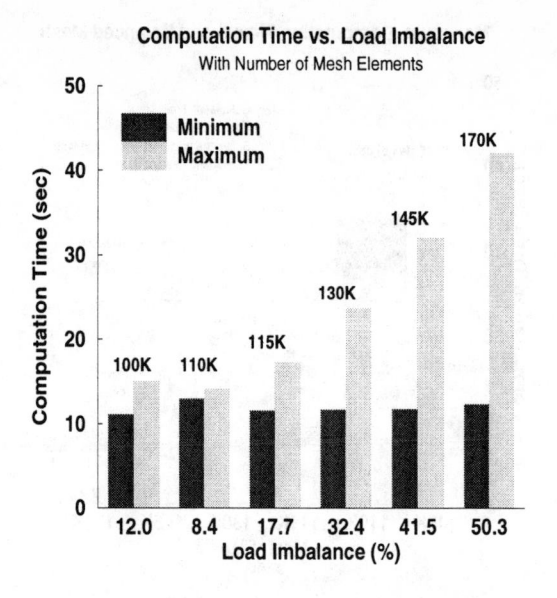

Fig. 4. Parallel CDT mesh generation performance: min/max computation time, no load balancing

ration of concerns philosophy; the DMCS layer maintains the correct message ordering, while the MOL maintains the causality of messages. User code would be responsible for the migration of objects and for the maintainance of the validity of references to those objects, if it relied solely upon a message passing system, especially with one-sided communication protocols.

At the other extreme of the continuum are page-based software DSM systems, in which a specific range of virtual memory is actually shared among all nodes in the parallel system. The MOL differs from these systems, in that the MOL needs no hardware, operating system, or compiler support. The MOL is implemented as a library that is linked with user code to provide some of the functionality of a DSM system. However, because it is a library, no complex interactions with low-level software or hardware are necessary; there is no interaction with the virtual memory system, the operating system, or physical memory busses. The trade-off for this simplicity is extra complexity of the programming model. Reads and writes to mobile objects do not utilize the same mechanism as do reads and writes to shared objects.

6 Summary and Conclusions

We have presented a run-time substrate to support the efficient data-migration required for the parallelization of adaptive applications. The runtime substrate automatically maintains the validity of global pointers as data migrates from one processor to another, and implements a correct and efficient message forwarding

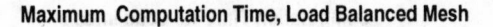

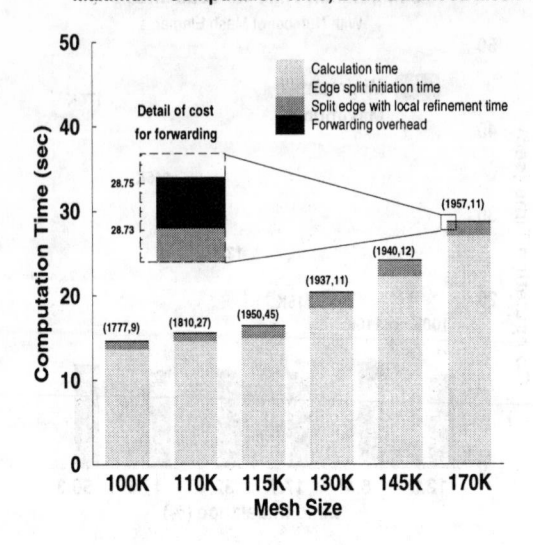

Fig. 5. Parallel CDT mesg generation performance: maximum computation time, load balancing

and communication mechanism between the migrating objects. The flexibility of our approach combined with the low overhead makes Mobile Object Layer attractive for application developers as well as compiler writers.

The MOL is specifically designed to reduce the amount of effort needed to efficiently and easily implement mobile, adaptive applications, and related support libraries. The most obvious benefit is that the programmer need not worry about the details of correct message passing in the presence of data migration, since it is handled within the MOL. Otherwise, several thousand lines of code would be written to perform similar functions to the MOL, and possibly less efficiently. Hence, more effort can be devoted to developing the application, instead of lower-level message passing and updating primitives.

The MOL is also lightweight, in that its latency is very close to that of the message layer it is built upon, even for forwarded messages. Thus, little is lost in the efficiency of an application relying upon the MOL to effect object migration. This is accomplished by only doing a minimal amount of extra computation to maintain the distributed directories and the incoming and outgoing message pools. However, the results of this minimalism show up in the MOL interface, which has a total of just six procedures. The current interface must be expanded and enhanced; a number of additions can be made to increase the capabilities of the MOL.

7 Future Work

To further improve the MOL's flexibility and efficiency, coherency protocols will be implemented as "plug and play" modules to maximize performance for a given application. Different applications benefit from different coherency models, so restricting the programmer to only a single coherency model unnecessarily degrades performance. By allowing user code to experiment with different protocols, the MOL can be tweaked to obtain the highest level of performance on an application by application basis.

All of the coherency protocols in use today fall into three categories: "lazy" updating, "eager" invalidation, and "eager" updating protocols. Each of these protocols results in a certain amount of network traffic under certain circumstances, and therefore the best choice for which coherency protocol to use can vary from one application to another. For example, in situations where nodes communicate infrequently with migrated mobile objects, a lazy updating protocol, such as the one that is currently implemented in the MOL, works better relative to an eager update or invalidation protocol. A lazy protocol avoids unnecessary broadcasting of update or invalidation messages to nodes which no longer have an interest in a mobile object.

However, if an eager protocol has hardware support, a software-based lazy protocol may not be the best choice. For example, in the FLASH [21] system, the network controllers are programmed to implement eager protocols. In the explicit updating protocol used by FLASH, all nodes that hold pointers to a shared memory region are automatically updated if the memory is invalidated.

Finally, we will also change the dense, two-dimensional array directory structure into a sparse data structure, such as a hashtable. A matrix data structure is non-scalable with respect to memory usage, although it supports very fast retrieval of mobile objects. Careful implementation of the sparse data structure will provide access speeds comparable to that of the matrix data structure.

By affording programmers the ability to choose the coherency protocol, the MOL can be tailored to work more cohesively with the application. Support for multiple message sizes will also further this goal, and, along with replacing the dense directory structure with a sparse structure, will reduce the memory overhead imposed upon applications which utilize the MOL for message passing and object mobility.

Acknowledgements We would like to thank the referees for their constructive comments.

References

[1] Chrisochoides, N., Pingali, K., Kodukula, I.; Data movement and control substrate for parallel scientific computing *Lecture Notes in Computer Science (LNCS), Springer-Verlag* **1199**, *256* 1997.

[2] Foster, I., Kesselamn, C., Tuecke, S.; The Nexus task-parallel runtime system, In *Proc. 1st Int. Workshop on Parallel Processing*, 1994.

[3] Beckman, P., Gannon, D.; Tulip: Parallel run-time support system for pC++ (1996), http://www.extreme.indiana.edu.

[4] DiNicola, P., Gildea, K., Govindaraju, R., Mirza, J., Shah, G.; LAPI architecture definition: Low level API draft, IBM Confidential, December 1996.

[5] Hawblitzel, C., Chrisochoides, N.; Mobile object layer: A data migration framework for Active Messages paradigm, University of Notre Dame Department of Computer Science and Engineering TR 98-07, 1998.

[6] Fowler, R.; The complexity of using forwarding addresses for decentralized object finding, In *Proc. 5th Annual ACM Symp. on Principles of Distributed Computing*, 1986.

[7] von Eicken, T., Culler, D., Goldstein, S., Schauser, K.; Active messages: A mechanism for integrated communication and computation, In *Proc. 19th Int. Symp. on Computer Architecture*, 1992.

[8] Blumofe, R., Leiserson, C.; Scheduling multithreaded computations by work stealing, In *FOCS-35*, pp. 356-368, 1994.

[9] Chrisochoides, N.; Multithreaded model for load balancing parallel, adaptive computations on multicomputers, *J. Appl. Num. Math*, **6** (1996), pp. 1-17.

[10] Chew, L. Paul, Chrisochoides, N., Sukup, F.; Parallel constrained Delaunay meshing, In *Proc. Joint ASME/ASCE/SES Summer Meeting, Special Symp. on Trends in Unstructured Mesh Generation*, 1997.

[11] Chew, L. Paul; Constrained Delaunay triangulations, *Algorithmica*, **4** (1989), 97–108.

[12] Chrisochoides, N., Nave, D., Hawblitzel, C.; Data migration substrate for the load balancing of parallel adaptive unstructured mesh computations, In *Proc. 6th Int. Conf. on Numerical Grid Generation in Computational Field Simulation*, 1998.

[13] Chang, C., Czajkowski, G., Hawblitzel, C., von Eicken, T.; Low-latency communication on the IBM RISC System/6000 SP, In *Proc. SC '96*, 1996.

[14] Arjomandi, E., O'Farrell, W., Kalas, I., Koblents, G., Eigler, F., Gao, G.; ABC++: Concurrency by inheritance in C++, IBM Sys. J., Vol. 34, No.1, 1995, pp. 120-137.

[15] Amza, C., Cox, A., Dwarkadas, S., Keleher, P., Lu, H., Rajamony, R., Yu, W., Zwaenepoel, W.; TreadMarks: Shared memory computing on networks of workstations, (1996) *IEEE Computer*, **29(2)**, *18*.

[16] Kale, L., Krishnan, S.; "Charm++," in *Parallel Programming Using C++*, eds. Wilson, G. and Lu, P., The MIT Press, 1998.

[17] Chang, C., Sussman, A., Saltz, J.; "Chaos++", *Parallel Programming Using C++*, eds. Wilson, G. and Lu, P., The MIT Press, 1998.

[18] Kesselman, C.; "CC++," In *Parallel Programming Using C++*, eds. Wilson, G. and Lu, P., The MIT Press, 1998.

[19] Chase, J., Amador, F., Lazowska, E., Levy, H., Littlefield, R.; The Amber system: Parallel programming on a network of multiprocessors, SOSP-12, pp. 147-158, December, 1989.

[20] Johnson, K., Kaashoek, F., Wallach, D.; CRL: High-performance all-software distributed shared memory, In *Proc. 15th Annual Symp. on OS Principles*, 1995.

[21] Kuskin, J., Ofelt, D., Heinrich, M., Heinlein, J., Simoni, R., Gharachorloo, K., Chapin, J., Nakahira, D., Baxter, J., Horowitz, M., Gupta, A., Rosenblum, M., Hennessy, J.; The Stanford FLASH multiprocessor, In *Proc. 21st Int. Symp. on Computer Architecture*, 1994.

Software Tools for Partitioning Block-Structured Applications

Jarmo Rantakokko

Uppsala University, Uppsala, Sweden

Abstract. A flexible software package for data partitioning has been developed. The package considers irregularly weighted structured grids and irregularly coupled structured multiblock grids. But, also unstructured partitioning can be addressed with the software tools. The software gives support for construction of different partitioning algorithms by composition of low-level operations. Automatic partitioning methods are also included. The implementation is in Fortran 90 with an object-oriented design. The use of the package has been demonstrated by partitioning a grid for an oceanographic model and a multiblock grid modeling an expanding and contracting tube for airflow computations.

1 Introduction

Structured grids are commonly used in scientific computing, e.g. in airflow simulations, ocean modeling and electro-magnetic computations. For complicated geometries a set of block-structured grids, i.e. a composite grid, is needed. An alternative is to use unstructured grids. However, structured grids require less memory and efficient solving techniques can be more easily implemented [1, 2]. For unstructured grid problems there are general partitioning methods that work well for many applications and a number of general software packages are available for these methods, for example *Top/Domdec* [3], *Chaco* [4], *Metis* [5], *Jostle* [6], and *Scotch* [7]. For structured grid problems we also have the constraint that the partitioning methods should yield structured partitions in order to preserve the efficiency of the solvers. Then, there is a trade-off between the structure and the load balance which depends very much on the application. Consequently, the software for partitioning is usually integrated in the parallel solver environment and can not easily be extracted and adapted to other problems or applications.

Partitioning a single grid with a homogeneous workload is straightforward. The grid can simply be divided into equally sized rectangular blocks, one block per processor. This can easily be handled with a data parallel compiler, for example High Performance Fortran (HPF) [8]. The problem arises when the workload becomes irregular or we have irregular data dependencies. For a single structured grid with an irregular workload in the domain there exist a number of block partitioning algorithms [9], e.g. the Recursive Coordinate Bisection Method which is perhaps the most known. However, these standard methods do not always give satisfactory results, as discussed in Sect. 3.2, and we have developed a new approach to partition data for this class of problems. The idea is to compose an algorithm from a set of low-level operations. This will be the key issue for the software package discussed below. For composite grid problems, i.e. irregularly coupled regular grids, the subgrids are usually partitioned either over all processors one by one or distributed as they are to different processors. The success of these two approaches is limited and they work well only for certain applications. We have also developed a new partitioning strategy. Our approach is very flexible and give good results for different kinds of composite grids. The ideas originate from our partitioning approach for single structured grids.

These two classes of problems, single structured grids with irregular workload and composite structured grids, are very much related and it appears that they can be solved with related algorithms and software. The partitioning software we describe in this paper address this issue and give support for previous strategies as well as the new algorithms proposed by us. The emphasis is to give an overview of the software and its capabilities. The strategies and algorithms are more thoroughly described elsewhere [10]. The design of the package is very important for the flexibility. The software is written in Fortran 90 with an object oriented-design and is an independent package that can easily be used in various applications. The object-oriented design provides a new way to construct partitioning algorithms by composition of low-level operations. The result is that we have a software package that is flexible enough to address a large variety of applications. The user can choose a suitable algorithm or even compose a new one from the low-level building blocks.

The rest of the paper has the following outline. The software package and the ideas are presented in Sect. 2. Then in Sect. 3, we briefly describe the kind of partitioning strategies that can be addressed with our software. In Sect. 4 we discuss some applications and finally in Sect. 5 we summarize our contributions.

2 The Software

The partitioning software is implemented in Fortran 90. It has an object-oriented design. This means that we have created abstract data types and encapsulated both the data and functionality into one entity, a class. The data can only be affected through the predefined operations within the class. This yields a disciplined way to program and use the software tools. But, it also gives a natural and flexible way to create partitioning algorithms by composition of the predefined operations. Even though Fortran 90 is not a fully object-oriented language it gives good support for object-oriented programming with the *type*, *module* and *interface* concepts [11]. Inheritance is not directly supported so it can be argued that the Fortran 90 implementation remains *object-based*.

We have three classes that address different kinds of problems, (i) *graph* for graph partitioning problems or unstructured grids, (ii) *domain decomposition (dd)* for structured irregular workload problems, and (iii) *composite domain decomposition (cdd)* for irregularly coupled structured grids. In addition, we have a corresponding set of derived composite data types to facilitate the communication with the classes giving cleaner interfaces. The user only has to understand how these derived data types are constructed to adapt the partitioning software to his or her application. The software infrastructure is illustrated in Fig. 1.

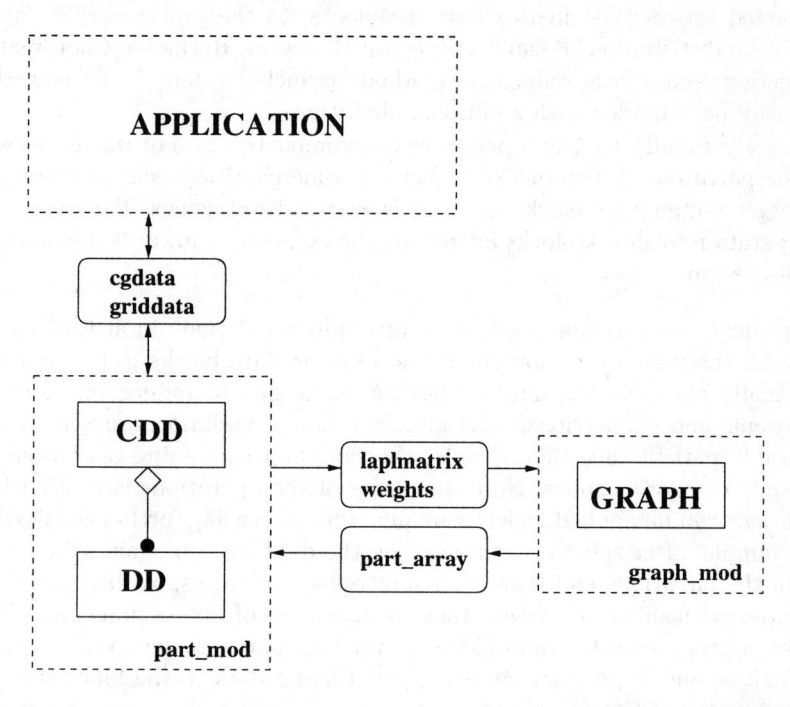

Fig. 1. System infrastructure. We have a module (dashed frame) for block-structured partitioning, *part_mod*, and a module for graph partitioning, *graph_mod*. The separate modules communicate through simple communication objects, *cgdata, griddata, laplmatrix, weights,* and *part_array* which also are encapsulated in their own module

The *graph* class contains operations for partitioning a graph. Here, we have implemented the Recursive Spectral Bisection method, a Greedy-like partitioning method, and a local refinement method as operations in the class. A graph is created from a Laplacian matrix and the vertex weights. As a result we can access the partitioning of the vertices. This is a separate and independent module that can be used directly in an application.

The class *domain decomposition* contains a set of low-level operations to partition and decompose an irregularly weighted domain into a set of structured data blocks. The operations in this class can be divided into three categories:

Splitting: Here we have implemented methods to split the domain into either equal sized parts or equal weighted parts. We have also an operation that recursively splits blocks depending on their relative sizes or weights. In addition, we allow the user to define its own splitting of the domain. Then the user can easily incorporate her favorite partitioning algorithm into our framework.

Mapping: This category includes operations to map the data to the processors. We have implemented various partitioning methods that distributes a connected set of data, in this case the blocks, to the processors. So far only Cyclic distribution, Recursive Spectral Bisection, Recursive Coordinate Bisection, and a local refinement method are included but the framework can easily be extended with additional algorithms.

Assembly: Finally, we have operations to assemble the data or the blocks within the partitions. If two blocks within the same partitions can be merged to a larger rectangular block then this is more advantageous. We have also an operation to shrink blocks by moving the boundaries inwards if possible (see discussion in Sect. 3.2).

The principle to partition a grid is to first split the domain in at least as many blocks as there are processors, map the resulting data blocks to the processors, and finally assemble the data within the partitions to reduce the number of data items and connections. This gives the user a variation range to compose a suitable partitioning algorithm by choosing and combining operations from the three categories above. Note, the order of the operations is not fixed. The operations can be applied logically in any order increasing further the flexibility. For example, after splitting and mapping the data we can again split the data within the partitions and then re-map or refine the corresponding partitions in a multi-level fashion. Moreover, there is also a set of access functions. We can extract a graph with the data blocks as vertices, partition the graph within the *graph* class, and impose the resulting partitioning back to the block structured decomposition of the domain.

The class *composite domain decomposition* is an aggregate of several instances of the class *dd*, one for each subgrid. This means that we can access an individual subgrid and partition by using the methods in the class *dd*, i.e. partition the subgrids one by one using the single grid strategies. The *cdd* class also has an overview of the whole composite grid and handles the connections between the subgrids. For example we have an algorithm to cluster the processors, one processor cluster for each subgrid. We can then partition the subgrids in the different clusters independently using any single grid partitioning strategy. The idea is similar to the concept of processor subsets in HPF-2 but here we have developed an optimal clustering algorithm with respect to the load balance between the clusters. Moreover, we can create a connectivity graph with data items from all subgrids (considering the inter-grid connections). The data items or vertices in the graph can be the subgrids, the data blocks from splitting the grids, or even the individual grid points in the subgrids. This gives us lots of flexibility and we can compose partitioning algorithms ranging from a structured coarse grain parallelism down to an unstructured fine grain parallelism.

For evaluation and post-processing purposes, the three classes generate statistics of the corresponding partitionings. This can also be used to create automatic and adaptive partitioning algorithms. Furthermore, data files are produced for visualization in Matlab. (The Matlab post-processing scripts are included in the software package.)

In summary, the *dd* class is the key component in the system. It includes low-level operations to create block-structured partitions. A composite domain can be represented and partitioned using a set of *dd*-objects. The *cdd* class has an overview of the composite grid and includes in addition some global partitioning operations. The *graph* class is an auxiliary class for the blocks-structured partitioning applications. The separation makes it easy to incorporate other graph partitioning methods or to connect other graph partitioning packages to our system, such as *Chaco* or *Metis*. Only the code in the graph-module has to be adapted.

3 Partitioning Strategies and Methods

This section is a short overview of partitioning algorithms for the kind of problems we address with our software. The section also serves to give some additional indication on how our tools are intended to be used. We have three subsections but the emphasis is on the two latter for the block-structured applications.

3.1 Graph Partitioning

In graph partitioning a weighted graph is constructed with the data items as vertices. The edges in the graph correspond to the neighbor relations between the data items. The graph can be represented numerically with a *Laplacian matrix*. The graph is then split in subgraphs with a partitioning method giving the corresponding data partitioning. The goal is to get compact subgraphs with a minimal number of edge-cuts.

The *Recursive Spectral Bisection method* [12] partitions the graph by computing the eigenvector corresponding to the second largest eigenvalue. This method is considered to give globally good partitioning results, but is computationally very demanding and the partitioning can be suboptimal in the fine details. The quality of the partitioning can often be improved with a *local refinement algorithm*. A local refinement method is also one building block in a multi-level method [13]. Multi-level methods decrease the complexity in the partitioning considerably and give often better results than the corresponding non-hierarchical methods. The vertex weights usually corresponds to the arithmetic work and can be very different for the different vertices. The load balance can then be considerably improved by ignoring the neighbor relations and by using some bin-packing method for the partitioning, e.g. variants of the *Greedy partitioning method* [14]. These algorithms are also very fast.

The three methods above are included in our graph-module. These operations are used to map a connected set of data blocks, originated from the structured applications below, to the processors. But, they can also be used directly in an application if a Laplacian matrix is provided.

3.2 Irregular Workload Problems

We consider the problem of partitioning a rectangular array with an irregular workload. This problem arises for example in ocean modeling where the sea depth is inhomogeneous [15]. For structured grids it is necessary to have structured partitions, i.e. blocks, to keep the efficiency in the solver. Also, this is a natural parallelization strategy. The original solver can then be reused on the blocks. To get a good load balance, the blocks can have different sizes or several blocks can be assigned to the same processors. The load balance is essential but a number of other, sometimes conflicting, objectives should also be met. We must consider the following issues in the partitioning:

- The load should be as even as possible.
- The number of communication points should be as low as possible.
- There should be as few blocks as possible.
- The blocks should be as dense as possible.

The last item states that the blocks should contain as little area as possible where no computations are required. For example, we should minimize the "empty" points corresponding to land-points in ocean modeling.

Recursive Coordinate Bisection is a fast and robust method to partition a rectangular domain into equally weighted blocks, but does not consider the last item above. The whole domain is partitioned into equally weighted blocks and consequently all blocks contain work. If the domain is partitioned in a number of blocks without regarding the workload some blocks may be completely "empty" and can be removed, reducing the number of unused points. The load balance will then depend on the number of blocks and their sizes. Many small blocks give a good load balance but also an increased overhead in moving data between the blocks. A strategy, used for example in partitioning the Baltic sea [15], is to (i) divide the domain uniformly in a number of blocks, (ii) remove the empty blocks, (iii) split only the heavy blocks to smooth the workload between the blocks, (iv) map the resulting blocks to the processors using a graph partitioning method, (v) merge blocks within the partitions to larger rectangular blocks, and (vi) shrink the boundaries of the blocks if possible to reduce further the number of unused points. This approach includes the three types of operations, *splitting, mapping*, and *assembly*, which are supported in the partitioning software.

3.3 Composite Grids

The partitioning of composite grids is complicated by the irregular data dependencies between the subgrids and of the different subgrid sizes. The number of subgrids and their sizes can differ much in various applications. The partitioning strategy then becomes very dependent on the specific application.

Composite grids exhibit two levels of parallelism, between the grids and within the grids. Exploiting either the coarse-grain parallelism, i.e. the subgrid level, or the fine-grain parallelism, i.e. the grid point level within the subgrids, is straightforward and is commonly used in computational fluid dynamics (CFD) applications with multiblock grids. A further development of these two strategies is to divide the processors into subsets, one per subgrid, and to partition the grids within the different subsets of processors. The success of these methods depends very much on the number of subgrids and their sizes as well as the number of available processors.

We have developed a new approach [10] where we exploit both levels of the inherent parallelism in an efficient way. We consider all couplings between the subgrids and at the same time balance the arithmetic workload very well. The partitioning algorithm can be composed by low-level operations. The basic idea to partition the complete composite grid on all the processors is to divide each element grid into a number of smaller data blocks, set up a connectivity graph for the blocks and apply a graph partitioning method. Then, merge small blocks into larger rectangular blocks within each processor and subgrid. To get a general and efficient algorithm, the partitioning can be done in a multilevel fashion. The subgrids can be split recursively until the number of blocks is larger than the number of processors. Then the blocks can be mapped to the processors using a graph partitioning method. To get an even better load balance the blocks can be split further, projecting the partitioning, and refining with a local refinement method. Finally, the blocks can be merged within the processors.

The low-level operations are supported in the software package and the outlined algorithm can easily be implemented by composition of the different operations from the respective classes. In addition, the previous methods described above are available. This gives the user a freedom to choose the best possible algorithm from an asset of methods for her application.

4 Applications

With the software tools we can easily compose a partitioning algorithm that adapts to a specific application. As two realistic examples we have partitioned an oceanographic model for the Baltic sea and a multiblock grid used in computational fluid dynamics.

The Baltic sea application contains an irregular workload due to varying sea depth and a large fraction of land points in the computational grid, see Fig. 2. We have used the strategy described in Sect. 3.2 to partition the Baltic sea. Compared to some standard partitioning methods our *block* decomposition gives very promising results, see Table 1. The unstructured partitioning with the Recursive Spectral Bisection method is not applicable unless substantial recoding of the solver is made. The Recursive Coordinate Bisection method has a high fraction of land-filled grid points, more than 90%, and a large number of edge-cuts. The straightforward uniform partitioning has to much load imbalance to be efficient. Our new block method gives a good compromise between these different requirements.

The Swedish Meteorological and Hydrological Institute is currently in the process of parallelizing their operational model for the Baltic sea and will be using the described partitioning approach and our software tools. No actual runtimes are yet available for this problem.

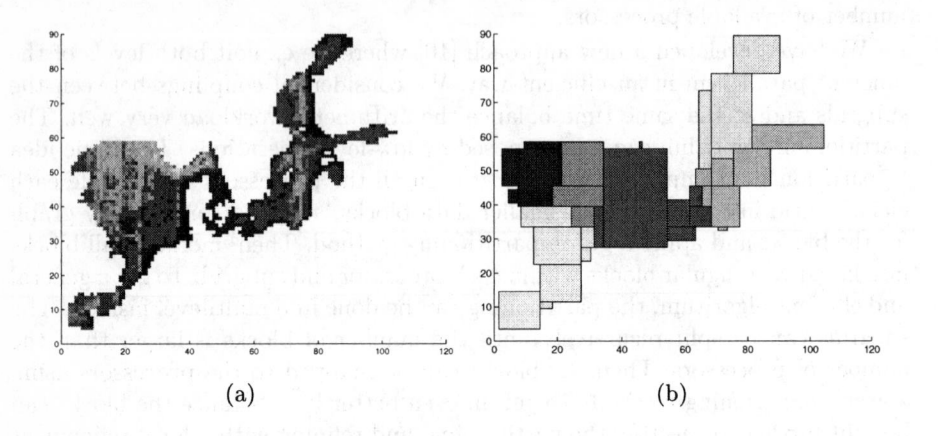

(a)   (b)

Fig. 2. The Baltic Sea (left) with block-structured partitioning (right). We have irregular workload due to varying sea depth and land-filled grid points

Table 1. Our method, *Block*, is compared with three standard methods, *Uniform* block distribution (same as the (BLOCK,BLOCK) distribution in HPF), Recursive Coordinate Bisection (RCB), and Recursive Spectral Bisection (RSB). Few blocks, a high fraction of active points, a low load imbalance ratio, and a small number of edge-cuts are preferable. However, these requirements are conflicting.

Method	Blocks	Active	Load	Edge-cut
Uniform	6	0.082	1.78	287
RCB	6	0.082	1.028	287
RSB	2171	1.00	1.0015	112
Block	32	0.42	1.042	130

The other application is a multiblock grid. We have five blocks modeling an expanding and contracting tube, see Fig. 3. We have solved the compressible Navier–Stokes equations for airflow in the tube. The code is parallelized with the Cogito software tools [16] and we have partitioned the grids with our new partitioning approach described in Sect. 3.3. The numerical experiments[1], see Fig. 4, show that our software tools produce partitionings that are good. We

---

[1] The numerical experiments were performed at the Edinburgh Parallel Computing Centre supported by the *Training and Research on Advanced Computing Systems* (TRACS) programme.

have almost 50% efficiency on the largest processor configuration. Still, this grid is quite small to run on all the 512 processors.

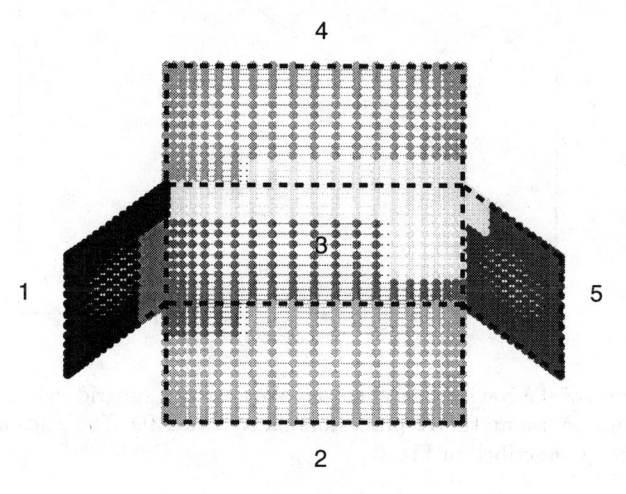

(a) Data partitioning

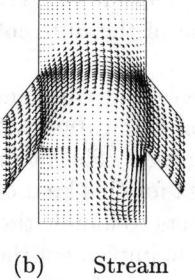

(b) Stream lines

Fig. 3. Simulation of airflow in an expanding and contracting tube. We have partitioned the multiblock grid by decomposing the individual subgrids in smaller data blocks, setting up a graph with the blocks as vertices, and partitioning the graph with the Recursive Spectral Bisection method

Our software gives support to implement the other strategies for partitioning composite grids as well but it will be difficult to compare them for the expanding and contracting tube case. A real-life application contains typically about ten times more subgrids than our example, e.g. the SAAB-2000 aircraft is modeled with 48 subgrids of different sizes and shapes. Therefore we have made the comparisons for a simpler application, the advection equations in 2D, but modeled the geometry with 20 subgrids. The results from the experiments are

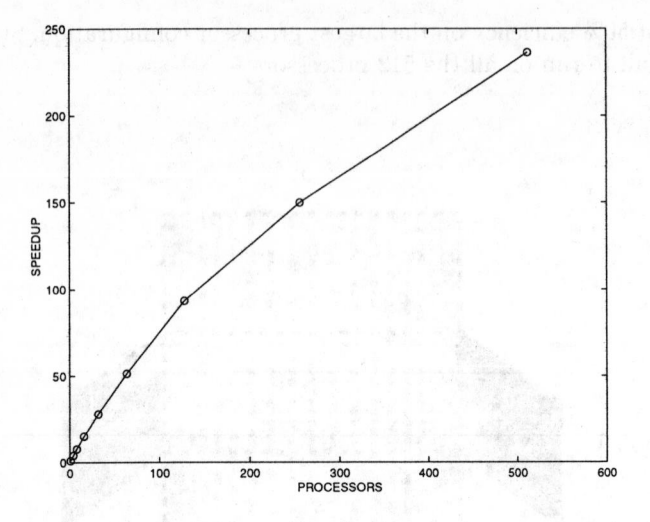

Fig. 4. Speedup of the Navier–Stokes solver. We have five subgrids within total 450,000 grid points and we use up to 512 processors on a Cray T3D. The grids are partitioned with the strategy described in Fig. 3

shown in Fig. 5. We can see that our method gives the best results except for large processor configurations where the clustering strategy is preferred. A more extensive study and comparison of the different partitioning strategies may be found elsewhere [17].

The implementation of solvers that can handle the kind of partitions shown in Fig. 3 is non-trivial. The communication will have an irregular pattern and special care must be taken to get an efficient implementation. The Cogito software tools are specially designed for this kind of applications and use MPI with persistent communications objects to handle the update of ghostcells. The HPF-2 specification includes support to implement the different partitioning strategies but compilers that include all the new features are not yet available. The KeLP infrastructure [18] can handle the irregular communication pattern for a single grid very efficiently. A cooperation is now initiated to develop abstractions in KeLP to give support for composite grids as well.

5   Conclusions

We have constructed a set of software tools in Fortran 90 with an object-oriented design. The object-oriented design yields a set of low-level operations that can be combined in any order to get a flexible partitioning algorithm. As an additional result, the partitioning framework also serves as a testbed to construct, evaluate, and visualize different partitioning algorithms.

The software tools address both unstructured graph partitioning and structured block partitioning. The emphasis is on the latter category. The applications

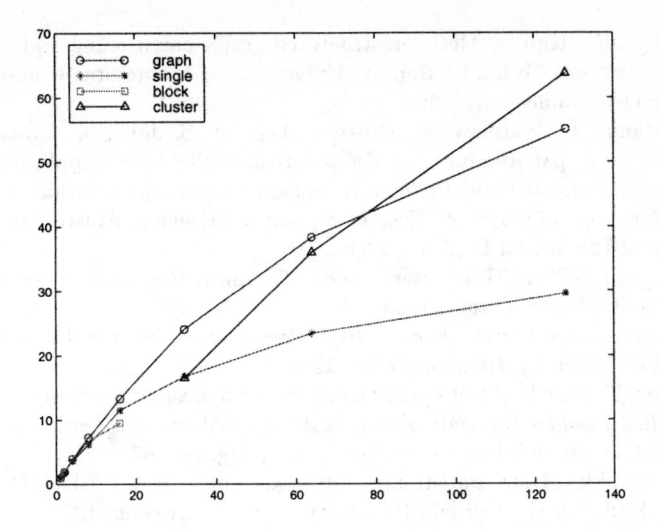

Fig. 5. Speedup of the advection equation solver. We have 20 subgrids within total 90,000 grid points and we use up to 128 processors on a Cray T3D. Four different partitioning strategies are compared, (1) *graph*: our new partitioning method, (2) *single*: partitioning the grids one at a time over all processors, (3) *block*: distributing the subgrids as they are to different processors, and (4) *cluster*: clustering the processors for the subgrids and partitioning the subgrids within the different processor clusters

are for example ocean modeling and multiblock grids within CFD. The partitioning software provides a new way to compose a partitioning algorithm with interaction from the user. Traditional software packages try to give a universal solution to the partitioning problem, which may not be optimal from application to application, or is only limited to a specific problem. Our package is general enough to be efficient for a large variety of partitioning problems. We have shown this by partitioning data for an oceanographic model and a multiblock grid used in airflow computations. For both of these applications we have been able to construct new partitioning algorithms with our software that in some cases are better than the previous standard algorithms.

References

[1] E. Pärt-Enander. *Overlapping Grids and Applications in Gas Dynamics*, Ph.D. thesis, Uppsala University, Uppsala, Sweden, 1995.
[2] E. Steinthorsson, D. Modiano. Advanced methodology for simulation of complex flows using structured grid systems. Report ICOMP-95-28, Institute for Computational Mechanics in Propulsion, Cleveland, Ohio, 1995.
[3] C. Farhat, S. Lanteri, H.D. Simon. TOP/DOMDEC - A software tool for mesh partitioning and parallel processing. *Computing Systems in Engineering*, Vol. 6, 1:13-26, 1995.
[4] B. Hendrickson, R. Leland. The Chaco user's guide, version 2.0. Technical Report, Sandia National Laboratories, Albuquerque, NM, July 1995.

[5] G. Karypis, V. Kumar. Metis: unstructured graph partitioning and sparse matrix ordering system. Technical Report, University of Minnesota, Computer Science Department, Minneapolis, 1995.

[6] K. McManus, C. Walshaw, M. Cross, P. Leggett, S. Johnson. Evaluation of the JOSTLE mesh partitioning code for practical multiphysics applications. In *Parallel Computational Fluid Dynamics: Implementations and Results Using Parallel Computers*, pp. 673–680, A. Ecer et al, editor, Elsevier, Amsterdam, 1996. Proceedings of Parallel CFD'95, Pasadena.

[7] F. Pellegrini. SCOTCH 3.1 user's guide. Technical Report 1137-96, LaBRI, University of Bordeaux, France, 1996.

[8] High Performance Fortran Forum. *High Performance Fortran Language Specification*, Rice University, Houston Texas, 1993.

[9] F. Manne, T. Sørvik. Partitioning an array onto a mesh of processors. In *Workshop on Applied Parallel Computing in Industrial Problems and Optimization*, Lecture Notes in Computer Science 1184, Springer-Verlag, pp 467–477.

[10] J. Rantakokko. Data Partitioning Methods and Parallel Block-Oriented PDE Solvers. Ph.D. thesis, Uppsala University, Uppsala, Sweden, 1998.

[11] C.D. Norton. Object-Oriented Programming Paradigms in Scientific Computing. Ph.D. thesis, Rensselaer Polytechnic Institute, Troy, New York, 1996.

[12] H.D. Simon. Partitioning of unstructured problems for parallel processing. *Computing Systems in Engineering*, 2:135–148, 1991.

[13] B. Hendrickson, R. Leland. A multilevel algorithm for partitioning graphs. In *Proc. Supercomputing '95*, 1995.

[14] M.A. Iqbal, J.H. Saltz, S.H. Bohkari. Performance tradeoffs in static and dynamic load balancing strategies. Technical Report 86-13, ICASE, NASA Langley Research Center, Hampton, VA, 1986.

[15] J. Rantakokko. A framework for partitioning structured grids with inhomogeneous workload. To appear in *Parallel Algorithms and Applications*.

[16] P. Olsson, J. Rantakokko, M. Thuné. Software tools for parallel CFD on composite grids. In *Parallel Computational Fluid Dynamics: Implementations and Results Using Parallel Computers*, pp. 725–732, A. Ecer et al, editor, Elsevier, Amsterdam, 1996.

[17] J. Rantakokko. Comparison of partitioning strategies for PDE solvers on multi-block grids. To appear in *Proc. PARA98 workshop on Applied and Parallel Computing in Large Scale Scientific and Industrial Problems*, Umeå, Sweden, June 14–17, 1998.

[18] S.J. Fink, S.R. Kohn, S.B. Baden. Flexible communication mechanism for dynamic structured applications. In *Proc. IRREGULAR'96*, pp. 203–215, Santa Barbara, CA.

An Object-Oriented Collection of Minimum Degree Algorithms[*]

Gary Kumfert[1] and Alex Pothen[1,2]

[1] Old Dominion University, Norfolk, VA, USA
[2] NASA Langley Research Center, Hampton, VA, USA

Abstract. The multiple minimum degree (MMD) algorithm and its variants have enjoyed more than 20 years of research and progress in generating fill-reducing orderings for sparse, symmetric, positive definite matrices. Although conceptually simple, efficient implementations of these algorithms are deceptively complex and highly specialized.

In this case study, we present an object-oriented library that implements several recent minimum degree-like algorithms. We discuss how object-oriented design forces us to decompose these algorithms in a different manner than earlier codes and demonstrate how this impacts the flexibility and efficiency of our C++ implementation. We compare the performance of our code against other implementations in C or Fortran.

1   Introduction

We have implemented a family of algorithms in scientific computing, traditionally written in Fortran 77 or C, using object-oriented techniques and C++. The particular family of algorithms chosen, the Multiple Minimum Degree (MMD) algorithm and its variants, is a fertile area of research and has been so for the last twenty years. Several significant advances have been published as recently as the last three years. Current implementations, unfortunately, tend to be specific to a single algorithm, are highly optimized, and are generally not readily extensible. Many are also not in the public domain.

Our goal was to construct an object-oriented library that provides a laboratory for creating and experimenting with these newer algorithms. In anticipation of new variations that are likely to be proposed in the future, we wanted the code to be extensible. The performance of the code must also be competitive with other implementations.

These algorithms generate permutations of large, sparse, symmetric matrices to control the work and storage required to factor that matrix. We explain the details of how work and storage for factorization of a matrix depends on the ordering in Sect. 2. This is formally stated as the *fill-minimization* problem. Also in Sect. 2, we review the Minimum Degree algorithm and its variants emphasizing

---

[*] This work was supported by National Science Foundation grants CCR-9412698 and DMS-9807172, by a GAANN fellowship from the Department of Education, and by NASA under Contract NAS1-19480

recent developments. In Sect. 3 we discuss the design of our library, fleshing out the primary objects and how they interact. We present our experimental results in Sect. 4, examining the quality of the orderings obtained with our codes, and comparing the speed of our library with other implementations. The exercise has led us to new insights into the nature of these algorithms. We provide some interpretation of the experience in Sect. 5.

2 Background

We illustrate the effect ordering has on the work and storage requirements of matrix factorization, translate this to a useful graph theoretic tool, explain the rationale in which heuristic algorithms attempt to control work and storage, and mention a specialized data structure common to all competitive Minimum Degree-like algorithms called the *quotient graph*.

2.1 Sparse Matrix Factorization

Consider a linear system of equations $Ax = b$, where the coefficient matrix A is sparse, symmetric, and either positive definite or indefinite. A direct method for solving this problem computes a factorization of the matrix $A = LBL^T$, where L is a lower triangular matrix, and B is a block diagonal matrix with 1×1 or 2×2 blocks.

The factor L is computed by setting $L_0 = A$ and then creating L_{k+1} by adding multiples of rows and columns of L_k to other rows and columns of L_k. This implies that L has nonzeros in all the same positions[1] as A plus some nonzeros in positions that were zero in A, but induced by the factorization. It is exactly these nonzeros that are called *fill* elements. The presence of fill increases both the storage and work required in the factorization.

An example matrix is provided in Fig. 1 that shows non-zeros in original positions of A as "×" and fill elements as "•". This example incurs two fill elements. The order in which the factorization takes place greatly influences the amount of fill. The matrix A is often permuted by rows and columns to reduce the number of fill elements, thereby reducing storage and flops required for factorization. Given the example in Fig. 1, the elimination order $\{2, 6, 1, 3, 4, 5\}$ produces only one fill element. This is the minimum number of fill elements for this example.

If A is positive definite, Cholesky factorization is numerically stable for any symmetric permutation of A, and the fill-reducing permutation need not be modified during factorization. If A is indefinite, then the initial permutation may have to be further modified during factorization for numerical stability.

---

[1] No "accidental" cancellations will occur during factorization if the numerical values in A are algebraic indeterminates.

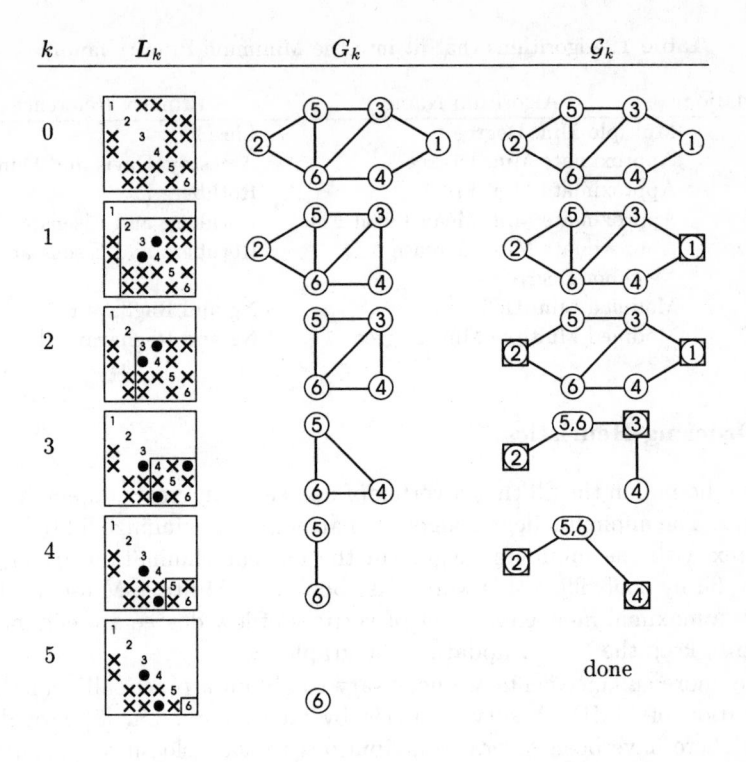

Fig. 1. Examples of factorization and fill. For each factorization step, k, there is the nonzero structure of the factor, L_k, the associated elimination graph, G_k, and the quotient graph $\mathcal{G}_k$. The elimination graph consists of vertices and edges. The quotient graph has edges and two kinds of vertices, supernodes (ovals) and enodes (boxed ovals).

2.2 Elimination Graph

The graph G of the sparse matrix A is a graph whose vertices correspond to the columns of A. We label the vertices $1, \ldots, n$, to correspond to the n columns of A. An edge (i, j) connecting vertices i and j in G exists if and only if a_{ij} is nonzero. By symmetry, $a_{j,i}$ is also nonzero.

The graph model of symmetric Gaussian elimination was introduced by Parter [1]. A sequence of elimination graphs, G_k, represent the fill created in each step of the factorization. The initial elimination graph is the graph of the matrix, $G_0 = G(A)$. At each step k, let v_k be the vertex corresponding to the kth column of A to be eliminated. The elimination graph at the next step, G_{k+1}, is obtained by adding edges to make all the vertices adjacent to v_k pairwise adjacent to each other, and then removing v_k and all edges incident on v_k. The inserted edges are *fill edges* in the elimination graph. This process repeats until all the vertices are removed from the elimination graph. The example in Fig. 1 illustrates the graph model of elimination. Finding an elimination order that produces the minimum amount of fill is NP-complete [2].

Table 1. Algorithms that fit into the Minimum Priority family

Abbreviation	Algorithm Name	Primary Reference
MMD	Multiple Min. Degree	Liu [3]
AMD	Approximate Min. Degree	Amestoy, Davis and Duff [4]
AMF	Approximate Min. Fill	Rothberg [5]
AMMF	Approximate Min. Mean Local Fill	Rothberg and Eisenstat [6]
AMIND	Approximate Min. Increase in Neighbor Degree	Rothberg and Eisenstat [6]
MMDF	Modified Min. Deficiency	Ng and Raghavan [7]
MMMD	Modified Multiple Min. Degree	Ng and Raghavan [7]

2.3 Ordering Heuristics

An upper bound on the fill that a vertex of degree d can create on elimination is $d(d-1)/2$. The minimum degree algorithm attempts to minimize fill by choosing the vertex with the minimum degree in the current elimination graph, hence reducing fill by controlling this worst-case bound. In Multiple Minimum Degree (MMD), a maximal independent set of vertices of low degree are eliminated in one step to keep the cost of updating the graph low.

Many more enhancements are necessary to obtain a practically efficient implementation of MMD. A survey article by George and Liu [8] provides the details. There have been several contributions to the field since the survey. A list of algorithms that we implement in our library and references are in Table 1. Most of these adaptations increase the runtime by 5-25% but reduce the amount of arithmetic required to generate the factor by 10-25%.

2.4 The Quotient Graph

Up to this point we have been discussing the elimination graph to model fill in a minimum priority ordering. While it is an important conceptual tool, it has difficulties in implementation arising from the fact that the storage required can grow like the size of the factor and cannot be predetermined. In practice, implementations use a *quotient graph*, $\mathcal{G}$, to represent the elimination graph in no more space than that of the initial graph $G(\boldsymbol{A})$. A quotient graph can have the same interface as an elimination graph, but it must handle internal data differently, essentially through an extra level of indirection.

The quotient graph has two distinct kinds of vertices: *supernodes* and *enodes*[2] A supernode represents a set of one or more uneliminated columns of $\boldsymbol{A}$. Similarly, an enode represents a set of one or more eliminated columns of $\boldsymbol{A}$. The initial graph, $\mathcal{G}_0$, consists entirely of supernodes and no enodes; further, each supernode contains one column. Edges are constructed the same as in the elimination graph. The initial quotient graph, $\mathcal{G}_0$, is identical to the initial elimination graph, G_0.

---

[2] Also called "eliminated supernode" or "element" elsewhere.

When a supernode is eliminated at some step, it is not removed from the quotient graph; instead, the supernode becomes an enode. Enodes indirectly represent the fill edges in the elimination graph. To demonstrate how, we first define a *reachable path* in the quotient graph as a path $(i, e_1, e_2, \ldots e_p, j)$, where i and j are supernodes in $\mathcal{G}_k$ and $e_1, e_2, \ldots e_p$ are enodes. Note that the number of enodes in the path can be zero. We also say that a pair of supernodes i, j is *reachable* in $\mathcal{G}_k$ if there exists a *reachable path* joining i and j. Since the number of enodes in the path can be zero, adjacency in $\mathcal{G}_k$ implies reachability in $\mathcal{G}_k$. If two supernodes i, j are reachable in the quotient graph $\mathcal{G}_k$, then the corresponding two vertices i, j in the elimination graph G_k are adjacent in G_k.

In practice, the quotient graph is aggressively optimized; all non-essential enodes, supernodes, and edges are deleted. Since we are only interested in paths through enodes, if two enodes are adjacent they are amalgamated into one. So in practice, the number of enodes in all reachable paths is limited to either zero or one. Alternatively, one can state that, in practice, the *reachable set* of a supernode is the union of its adjacent supernodes and all supernodes adjacent to its adjacent enodes. This amalgamation process is one way how some enodes come to represent more than their original eliminated column.

Supernodes are also amalgamated but with a different rationale. Two supernodes are *indistinguishable* if their reachable sets (including themselves) are identical. When this occurs, all but one of the indistinguishable supernodes can be removed from the graph. The remaining supernode keeps a list of all the columns of the supernodes compressed into it. When the remaining supernode is eliminated and becomes an enode, all its columns can be eliminated together. The search for indistinguishable supernodes can be done before eliminating a single supernode using graph compression [9]. More supernodes become indistinguishable as elimination proceeds. An exhaustive search for indistinguishable supernodes during elimination is prohibitively expensive, so it is often limited to supernodes with identical adjacency sets (assuming a self-edge) instead of identical reachable sets.

Edges between supernodes can be removed as elimination proceeds. When a pair of adjacent supernodes share a common enode, they are reachable through both the shared edge and the shared enode. In this case, the edge can be safely removed. This not only improves storage and speed, but allows tighter approximations to supernode degree as well.

Going once more to Fig. 1, we consider now the quotient graph. Initially, the elimination graph and quotient graph are identical. After the elimination of column 1, we see that supernode 1 is now an enode. Note that unlike the elimination graph, no edge was added between supernodes 3 and 4 since they are reachable through enode 1. After the elimination of column 2, we have removed an edge between supernodes 5 and 6. This was done because the edge was redundant; supernode 5 is reachable from 6 through enode 2. When we eliminate column 3, supernode 3 becomes an enode, it absorbs enode 1 (including its edge to supernode 4). Now enode 3 is adjacent to supernodes 4, 5 and 6. The fill edge between supernodes 4 and 5 is redundant and can be removed. At this point 4, 5, and

$k \leftarrow 0$

while $k < n$

 Let m be the minimum known degree, $\deg(x)$, of all $x \in \mathcal{G}_k$.

 while m is still the minimum known degree of all $x \in \mathcal{G}_k$

  Choose supernode x_k such that $\deg(x_k) = m$

  for all of the p columns represented by supernode x_k:

   Number columns $(k+1) \ldots (k+p)$.

  Form enode e_k from supernode x_k and all adjacent enodes.

  for all supernodes x adjacent to c_k:

   Label $\deg(x)$ as "unknown."

  $k \leftarrow k + p$

 for all supernodes x where $\deg(x)$ is unknown:

  Update lists of adjacent supernodes and enodes of x.

  Check for various QuotientGraph optimizations.

  Compute $\deg(x)$.

Fig. 2. The multiple minimum degree algorithm defined in terms of a quotient graph

6 are indistinguishable. However, since we cannot afford an exhaustive search, a quick search (by looking for identical adjacency lists) finds only supernodes 5 and 6 so they are merged to supernode $\{5, 6\}$. Then supernode 4 becomes an enode and absorbs enode 3. Finally supernode $\{5, 6\}$ is eliminated. The relative order between columns 5 and 6 has no effect on fill.

We show the Multiple Minimum Degree algorithm defined in terms of a quotient graph in Fig. 2. A single elimination Minimum Degree algorithm is similar, but executes the inner **while** loop only once. We point out that we have not provided an exhaustive accounting of quotient graph features and optimizations. Most of the time is spent in the last three lines Fig. 2, and often they are tightly intertwined in implementations.

3 Design

To provide a basis for comparison, we briefly discuss the design and implementation characteristics of MMD [3] and AMD [4]. Both implementations were written in Fortran77 using a procedural decomposition. They have no dynamic memory allocation and implement no abstract data types in the code besides arrays.

GENMMD is implemented in roughly 500 lines of executable source code with about 100 lines of comments. The main routine has 12 parameters in its calling sequence and uses four subroutines that roughly correspond to initialization, supernode elimination, quotient graph update/degree calculation, and finalization of the permutation vector. The code operates in a very tight footprint and will often use the same array for different data structures at the same time. The code has over 20 goto statements and can be difficult to follow.

```
        // Major Classes
        QuotientGraph* qgraph;
        BucketSorter* sorter;
        PriorityStrategy* priority;
        SuperNodeList* reachableSuperNodes, * mergedSuperNodes;

        // Initialization...
        ...
        // Load all vertices into sorter
1.      priority->computeAndInsert(priority::ALL_NODES, qgraph, sorter);
2.      if ( priority->requireSingleElimination() == true )
3.          maxStep = 1;
        else
4.          maxStep = graph->size();

        // Main loop
5.      while ( sorter->notEmpty() ) {
6.          int min = sorter->queryMinNonemptyBucket();
7.          int step = 0;
8.          while ( ( min == sorter->queryMinNonemptyBucket() &&
                      ( step < maxStep ) ) {
9.              int snode = sorter->removeItemFromBucket( min );
10.             qgraph->eliminateSupernode( snode );
                SuperNodeList* tempSuperNodes;
11.             tempSuperNodes = qgraph->queryReachableSet( snode );
12.             sorter->removeSuperNodes( tempSuperNodes );
13.             *reachableSuperNodes += *tempSuperNodes;
14.             ++step;
            }
15.         qgraph->update( reachableSuperNodes, mergedSuperNodes );
16.         sorter->removeSuperNodes( mergedSuperNodes );
17.         priority->computeAndInsert( reachableSuperNodes,
                    qgraph, sorter );
18.         mergedSuperNodes->resize( 0 );
19.         reachableSuperNodes->resize( 0 );
        }
```

Fig. 3. A general Minimum Priority Algorithm using the objects described in Fig. 4

– Quotient Graph
  1. Must provide a method for extracting the Reachable Set of a vertex.
  2. Be able to eliminate supernodes on demand.
  3. Should have a separate lazy update method for multiple elimination.
  4. Should provide lists of compressed vertices that can be ignored for the rest of the ordering algorithm.
  5. Must produce an elimination tree or permutation vector after all the vertices have been eliminated.
  6. Should allow const access to current graph for various Priority Strategies.
– Bucket Sorter
  1. Must remove an item from the smallest non-empty bucket in constant time.
  2. Must insert an item-key pair in constant time.
  3. Must remove an item by name from anywhere in constant time.
– Priority Strategy
  1. Must compute the new priority for each vertex in the list.
  2. Must insert the priority-vertex pairs into the Bucket Sorter.

Fig. 4. Three most important classes in a minimum priority ordering and some of their related requirements.

AMD has roughly 600 lines of executable source code which almost doubles when the extensive comments are included. It is implemented as a single routine with 16 calling parameters and no subroutine calls. It is generally well structured and documented. Manually touching up our f2c conversion, we were able to easily replace the 17 goto statements with while loops, and break and continue statements. This code is part of the commercial Harwell Subroutine Library, though we report results from an earlier version shared with us.

The three major classes in our implementation are shown in a basic outline in Fig. 4. Given these classes, we can describe our fourth object; the MinimumPriorityOrdering class that is responsible for directing the interactions of these other objects. The main method of this class (excluding details, debugging statements, tests, comments, etc.) is approximately the code fragment in Fig. 3. By far the most complicated (and expensive) part of the code is line 15 of Fig. 3 where the graph update occurs.

The most elegant feature of this implementation is that the PriorityStrategy object is an abstract base class. We have implemented several derived classes, each one implementing one of the algorithms in Table 1. Each derived class involves overriding two virtual functions (one of them trivial). The classes derived from PriorityStrategy average 50 lines of code each. This is an instance of the Strategy Pattern [10].

The trickiest part is providing enough access to the QuotientGraph for the PriorityStrategy to be useful and extensible, but to provide enough protection to keep the PriorityStrategy from corrupting the rather complicated state information in the QuotientGraph.

Because we want our library to be extensible, we have to provide the PriorityStrategy class access to the QuotientGraph. But we want to protect that access so that the QuotientGraph's sensitive and complicated internal workings are abstracted away and cannot be corrupted. We provided a full-fledged iterator class, called ReachableSetIterator, that encapsulated the details of the Quotient-Graph from the PriorityStrategy, making the interface indistinguishable from an EliminationGraph.

Unfortunately, the overhead of using these iterators to compute the priorities was too expensive. We rewrote the PriorityStrategy classes to access the QuotientGraph at a lower level ... traversing adjacency lists instead of reachable sets. This gave us the performance we needed, but had the unfortunate effect of increasing the coupling between classes. However, the ReachableSetIterator was left in the code for ease of prototyping.

Currently we have implemented a PriorityStrategy class for all of the algorithms listed in Table 1. They all compute their priority as a function of either the *external degree*, or a tight *approximate degree*, of a supernode. Computing the external degree is more expensive, but allows multiple elimination. For technical reasons, to get the approximate degree tight enough the quotient graph must be updated after every supernode is eliminated, hence all algorithms that use approximate degree are single elimination algorithms[3]. For this reason, all previous implementations are either multiple elimination codes or single elimination codes, not both. The quotient graph update is the most complicated part of the code and single elimination updates are different from multiple elimination updates.

The MinimumPriorityOrdering class queries the PriorityStrategy whether it requires quotient graph updates after each elimination or not. It then relays this information to the QuotientGraph class which has different optimized update methods for single elimination and multiple elimination. The QuotientGraph class can compute partial values for external degree or approximate degree as a side-effect of the particular update method.

Given this framework, it is possible to modify the MinimumPriorityOrdering class to switch algorithms during elimination. For example, one could use MMD at first to create a lot of enodes fast, then switch to AMD when the quotient graph becomes more tightly connected and independent sets of vertices to eliminate are small. There are other plausible combinations because different algorithms in Table 1 prefer vertices with different topological properties. It is possible that the topological properties of the optimal vertex to eliminate changes as elimination progresses.

4 Results

We compare actual execution times of our implementation to an f2c conversion of the GENMMD code by Liu [3]. This is currently among the most widely

---

[3] Readers are cautioned that algorithms in Table 1 that approximate quantities other than degree could be multiple elimination algorithms. Rothberg and Eisenstat [6] have defined their algorithms using either external degree (multiple elimination) or approximate degree (single elimination).

Table 2. Comparison of quality of various priority policies. Quality of the ordering here is measured in terms of the amount of work to factor the matrix with the given ordering. Refer to Table 1 for algorithm names and references

		Work		Work (normalized)				
	problem	MMD	AMD	AMF	AMMF	AMIND	MMDF	MMMD
1.	commanche	1.76e+06	1.00	.89	.87	.87	.92	.89
2.	barth4	4.12e+06	1.00	.89	.83	.82	.86	.82
3.	barth	4.55e+06	1.02	.90	.84	.85	.91	.89
4.	ford1	1.67e+07	.98	.84	.87	.82	.89	.86
5.	ken13	1.84e+07	1.01	.89	.88	.96	.83	.87
6.	barth5	1.96e+07	1.00	.90	.81	.82	.72	.83
7.	shuttle_eddy	2.76e+07	.97	.87	.74	.74	.75	.81
8.	bcsstk18	1.37e+08	.98	.77	.78	.74	.86	.83
9.	bcsstk16	1.56e+08	1.02	.81	.84	.82	.82	.81
10.	bcsstk23	1.56e+08	.95	.79	.73	.75	.80	.81
11.	bcsstk15	1.74e+08	.97	.89	.84	.81	.84	.86
12.	bcsstk17	2.22e+08	1.10	.89	.85	.88	1.02	.89
13.	pwt	2.43e+08	1.03	.92	.87	.90	.88	.90
14.	ford2	3.19e+08	1.03	.76	.72	.70	.77	.77
15.	bcsstk30	9.12e+08	1.01	.97	.82	.79	.88	.87
16.	tandem_vtx	1.04e+09	.97	.77	.56	.66	.70	.77
17.	pds10	1.04e+09	.90	.88	.91	.87	.88	1.00
18.	copter1	1.33e+09	.96	.82	.62	.71	.79	.87
19.	bcsstk31	2.57e+09	1.00	.95	.67	.71	.94	.87
20.	nasasrb	5.47e+09	.95	.82	.70	.79	.93	.82
21.	skirt	6.04e+09	1.11	.83	.90	.76	.88	.83
22.	tandem_dual	8.54e+09	.97	.42	.51	.62	.72	.72
23.	onera_dual	9.69e+09	1.03	.70	.48	.57	.65	.71
24.	copter2	1.35e+09	.97	.73	.50	.61	.66	.69
	geometric mean		1.00	.84	.74	.77	.83	.83
	median		1.00	.85	.82	.79	.85	.83

used implementations. In general, our object-oriented implementation is within a factor of 3-4 of GENMMD. We expect this to get closer to a factor of 2-3 as the code matures. We normalize the execution time of our implementation to GENMMD and present them in Table 3. For direct comparison, pre-compressing the graph was disabled in our C++ code. We also show how our code performs with compression.

All runtimes are from a Sun UltraSPARC-5 with 64MB of main memory. The software was compiled with GNU C++ version 2.8.1 with the -0, and -fno-exceptions flags set. The list of 24 problems are sorted in nondecreasing order of the work in computing the factor with the MMD ordering. The numbers presented are the average of eleven runs with different seeds to the random number generator. Because these algorithms are extremely sensitive to tie-breaking, it is common to randomize the graph before computing the ordering.

Table 3. Relative performance of our implementation of MMD (both with and without precompression) to GENMMD. GENMMD does not have precompression. The problems are sorted in nondecreasing size of the Cholesky factor

	problem	$|V|$	$|E|$	time (seconds) GENMMD no compr.	time(normalized) C++ no compr.	compr.
1.	commanche	7,920	11,880	.08	5.88	5.81
2.	barth4	6,019	17,473	.06	6.17	6.42
3.	barth	6,691	19,748	.10	5.00	5.36
4.	ford1	18,728	41,424	.30	4.57	4.69
5.	ken13	28,632	66,486	3.61	.92	.94
6.	barth5	15,606	45,878	.28	4.96	4.97
7.	shuttle_eddy	10,429	46,585	.09	9.44	9.33
8.	bcsstk18	11,948	68,571	.44	4.59	4.89
9.	bcsstk16	4,884	142,747	.16	8.19	1.74
10.	bcsstk23	3,134	21,022	.22	4.32	4.34
11.	bcsstk15	3,948	56,934	.22	4.77	4.62
12.	bcsstk17	10,974	208,838	.30	5.97	2.33
13.	pwt	36,519	144,794	.58	6.16	6.32
14.	ford2	100,196	222,246	2.44	3.84	3.90
15.	bcsstk30	28,924	1,007,284	.95	5.79	1.67
16.	tandem_vtx	18,454	117,448	.85	4.11	4.13
17.	pds10	16,558	66,550	107.81	1.24	1.16
18.	copter1	17,222	96,921	.67	6.22	6.52
19.	bcsstk31	35,588	572,914	1.50	4.83	2.58
20.	nasasrb	54,870	1,311,227	2.06	6.14	2.44
21.	skirt	45,361	1,268,228	2.03	6.38	1.72
22.	tandem_dual	94,069	183,212	4.50	3.70	3.67
23.	onera_dual	85,567	166,817	4.23	3.65	3.69
24.	copter2	55,476	352,238	3.96	4.57	4.70
	geometric mean				4.61	3.53
	median				4.90	4.24

We refer the reader to Table 2 for relative quality of orderings and execution times. As with the previous table, the data represents the average of 11 runs with different seeds in the random number generator. The relative improvement in the quality of the orderings over MMD is comparable with the improvements reported by other authors, even though the test sets are not identical.

We have successfully compiled and used our code on Sun Solaris workstations using both SunPRO C++ version 4.2 and GNU C++ version 2.8.1.1. The code does not work on older versions of the same compilers. We have also compiled our code on Windows NT using Microsoft Visual C++ 5.0.

5 Conclusions

Contrary to popular belief, our implementation shows that the most expensive part of these minimum priority algorithms is *not* the degree computation, but the quotient graph update. With all other implementations—including GENMMD and AMD—the degree computation is tightly coupled with the quotient graph update, making it impossible to separate the costs of degree computation from graph update with any of the earlier procedural implementations. The priority computation (for minimum degree) involves traversing the adjacency set of each reachable supernode after updating the graph. Updating the graph, however, involves updating the adjacency sets of each supernode and enode adjacent to each reachable supernode. This update process often requires several passes.

By insisting on a flexible, extensible framework, we required more decoupling between the priority computation and graph update: between algorithm and data structure. In some cases, we had to increase the coupling between key classes to improve performance. We are generally satisfied with the performance of our code and with the value added by providing implementations of the full gamut of state-of-art algorithms. We will make the software publicly available.

Acknowledgements We thank Tim Davis and Joseph Liu for their help and insights from their implementations and experiences. We are especially grateful to Cleve Ashcraft for stimulating discussions about object-oriented design, efficiency, and programming tricks.

References

[1] S Parter. The use of planar graphs in Gaussian elimination. *SIAM Rev.*, 3:364–369, 1961.

[2] M. Yannakakis. Computing the minimum fill-in is NP-complete. *SIAM J. Algebraic and Discrete Methods*, pages 77–79, 1981.

[3] Joseph W. H. Liu. Modification of the minimum-degree algorithm by multiple elimination. *ACM Trans. on Math. Software*, 11:141–153, 1985.

[4] Patrick Amestoy, Timothy A. Davis, and Iain S. Duff. An approximate minimum degree ordering algorithm. *SIAM J. Matrix Anal. Appl.*, 17(4):886–905, 1996.

[5] Ed Rothberg. Ordering sparse matrices using approximate minimum local fill. Preprint, April 1996.

[6] Ed Rothberg and Stan Eisenstat. Node selection strategies for bottom-up sparse matrix ordering. *SIAM J. Matrix Anal. Appl.*, 19(3):682–695, 1998.

[7] Esmond G. Ng and Padma Raghavan. Performance of greedy ordering heuristics for sparse Cholesky factorization. Submitted to SIAM J. Mat. Anal. and Appl., 1997.

[8] Alan George and Joeseph W. H. Liu. The evolution of the minimum degree algorithm. *SIAM Rev.*, 31(1):1–19, 1989.

[9] Cleve Ashcraft. Compressed graphs and the minimum degree algorithm. *SIAM J. Sci. Comput.*, 16(6):1404–1411, 1995.

[10] Erich Gamma, Richard Helm, Ralph Johnson, and John Vlissides. *Design Patterns: Elements of Reusable Object-Oriented Software*. Addison Wesley Professional Computing Series. Addison Wesley Longman, 1995.

Optimizing Transformations of Stencil Operations for Parallel Object-Oriented Scientific Frameworks on Cache-Based Architectures*

Federico Bassetti, Kei Davis, and Dan Quinlan

Los Alamos National Laboratory, Los Alamos, NM, USA

Abstract. High-performance scientific computing relies increasingly on high-level, large-scale, object-oriented software frameworks to manage both algorithmic complexity and the complexities of parallelism: distributed data management, process management, inter-process communication, and load balancing. This encapsulation of data management, together with the prescribed semantics of a typical fundamental component of such object-oriented frameworks—a parallel or serial array class library—provides an opportunity for increasingly sophisticated compile-time optimization techniques. This paper describes two optimizing transformations suitable for certain classes of numerical algorithms, one for reducing the cost of inter-processor communication, and one for improving cache utilization; demonstrates and analyzes the resulting performance gains; and indicates how these transformations are being automated.

1 Introduction

Current ambitions and future plans for scientific applications, in part stimulated by the Accelerated Scientific Computing Initiative (ASCI), practically mandate the use of higher-level approaches to software development, particularly more powerful organizational and programming tools and paradigms for managing algorithmic complexity, making parallelism largely transparent, and more recently, implementing methods for code optimization that could not be reasonably expected of a conventional compiler.

An increasingly popular approach is the use of C++ object-oriented software *frameworks* or hierarchies of extensible libraries. The use of such frameworks has greatly simplified (in fact, made practicable) the development of complex serial and parallel scientific applications at Los Alamos National Laboratory (LANL) and elsewhere. Examples from LANL include Overture [1] and POOMA [2].

Concerns about performance, particularly relative to FORTRAN 77, are the single greatest impediment to widespread acceptance of such frameworks, and our (and others') ultimate goal is to produce FORTRAN 77 performance (or

---

* This work was performed under the auspices of the U.S. Department of Energy by Los Alamos National Laboratory under Contract No. W-7405-Eng-36.

better, in a sense described later) from the computationally intensive components of such C++ frameworks, namely their underlying *array classes*. There are three broad areas where potential performance, relative to theoretical machine capabilities, is lost: language implementation issues (which we address for C++ elsewhere [3]), communication, and with the trend toward ever-deeper memory hierarchies and the widening differences in processor and main-memory bandwidth, poor cache utilization.

Experience demonstrates that optimization of array class implementations themselves is not enough to achieve desired performance; rather, their *use* must also be optimized. One approach, championed by the POOMA project (and others), is the use of *expression templates* [4]. Another, being pursued by us, is the use of an optimizing preprocessor.

We present optimizing transformations applicable to stencil or stencil-like operations which can impose the dominant computational cost of numerical algorithms for solving PDEs. The first is a parallel optimization which hides communication latency. The second is a serial optimization which greatly improves cache utilization. These optimizations dovetail in that the first is required for the second to be of value in the parallel case. Last is an outline of an ongoing effort to automate these (and other) transformations in the context of parallel object-oriented scientific frameworks.

2 Array Classes

In scientific computing arrays are the fundamental data structure, and as such compilers attempt a large number of optimizations for their manipulation [5]. For the same reason, array class libraries are ubiquitous fundamental components of object-oriented frameworks. Examples include A++/P++ [6] in Overture, *valarray* in the C++ standard library [7], Template Numerical ToolKit (TNT) [8], the GNU Scientific Software Library (GNUSSL) [9], and an unnamed component of POOMA.

The target of transformation is the A++/P++ array class library which provides both serial and parallel array implementations. Transformation (and distribution) of A++/P++ array statements is practicable because, by design, they have no hidden or implicit loop dependence. Indeed, it is common to design array classes so that optimization is reasonably straightforward—this is clearly stated, for example, for *valarray* in the C++ standard library. Such statements vectorize well, but our focus is on cache-based architectures because they are increasingly common in both large- and small-scale parallel machines.

An example of an A++/P++ array statement implementing a stencil operation is

```
for (int n=0; n != N; n++)  // Outer iteration
    A(I) = (A(I-1,J) + A(I+1,J) + A(I,J-1) + A(I,J+1)) * 0.25;
```

The statement may represent either a serial or parallel implementation of Jacobi relaxation. In the parallel case the array data represented by A is distributed in some way across multiple processors and communication (to update the ghost boundary points along the edges of the partitioned data) is performed by the "=" operator. The syntax indicates that A denotes an array object of at least two dimensions, and I and J denote either one-dimensional *index* or *range* objects—sets or intervals (respectively) of indexes. Thus the loops over I and J are implicit, as is distribution and communication in the parallel case.

In this case the array is two-dimensional, with the first dimension ranging from 0 to $SIZE\_Y - 1$, and the second ranging from 0 to $SIZE\_X - 1$; and I and J denote range objects with indexes 1 through $SIZE\_Y - 2$ inclusive and 1 through $SIZE\_X - 2$ inclusive, respectively.

The equivalent (serial) C code is

```
for (int n=0; n!=N; n++) {  // Outer iteration
  for (int j=1; j!=SIZE_Y-1; j++)  //calculate new solution
    for (int i=1; i!=SIZE_X-1; i++)
      a_new[j][i] = (a[j][i-1] + a[j][i+1] +
                     a[j-1][i] + a[j+1][i]) * 0.25;
  for (int j=1; j!=SIZE_Y-1; j++)  //copy new to old
    for (int i=1; i!=SIZE_X-1; i++)
      a[j][i] = a_new[j][i];
}
```

3 Reducing Communication Overhead

Tests on a variety of multiprocessor configurations, including networks of workstations, shared memory, DSM, and distributed memory, show that the cost (in time) of passing a message of size N, cache effects aside, is accurately modeled by the function $L + CN$, where L is a constant per-message latency, and C is a cost per word. This suggests that *message aggregation*—lumping several messages into one—can improve performance.[1]

In the context of stencil-like operations, message aggregation may be achieved by widening the ghost cell widths. In detail, if the ghost cell width is increased to three, using A and B as defined before, A[0..99,0..52] resides on the first processor and A[0..99,48..99] on the second. To preserve the semantics of the stencil operation the second index on the first processor is 1 to 51 on the first pass, 1 to 50 on the second pass, and 1 to 49 on the third pass, and similarly on the second processor. Following three passes, three columns of A on the first processor must be updated from the second, and vice versa. This pattern of access is diagrammed in Fig. 1.

Clearly there is a tradeoff of computation for communication overhead. In real-world applications the arrays are often numerous but small, with communication time exceeding computation time, and the constant time L of a message exceeding the linear time CN. Experimental results for a range of problem sizes and number of processors is given in Figs. 2 and 3.

---

[1] Cache effects are important but are ignored in such simple models.

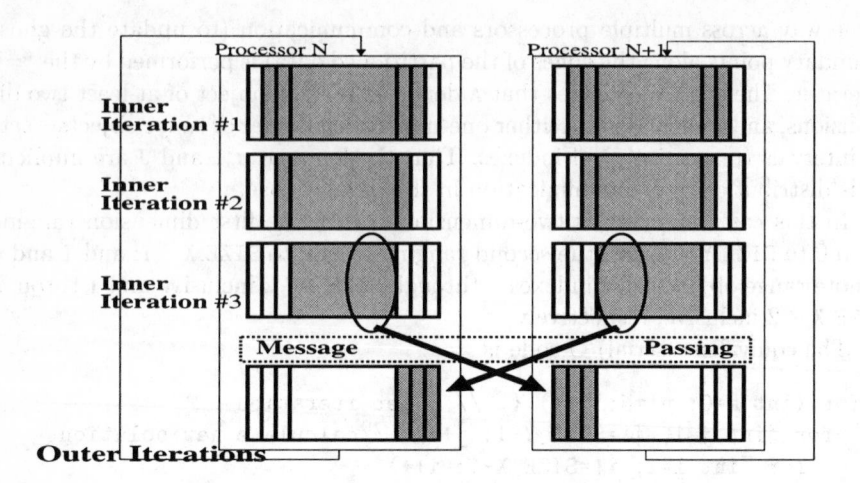

Fig. 1. Pattern of access and message passing for ghost boundary width three

Additional gains may be obtained by using asynchronous (non-blocking) message passing, which allows computation to overlap communication. Here the computation involving the ghost boundaries and adjacent columns is performed first, communication initiated, then interior calculations performed. Widening the ghost boundaries and so allowing multiple passes over the arrays without communication decreases the ratio of communication time to computation time; when the ratio is reduced to one or less communication time is almost entirely hidden.

4 Temporal Locality, Cache Reuse, and Cache Blocking

Temporal locality refers to the closeness in time, measured in the number of intervening memory references, between a given pair of memory references. Of concern is the temporal locality of references to the same memory location—if sufficiently local the second reference will be resolved by accessing cache rather than main or non-local shared memory.

A cache miss is *compulsory* when it results from the first reference to a particular memory location—no ordering of memory references can eliminate a compulsory miss. A *capacity* miss occurs when a subsequent reference is not resolved by the cache, presumably because it has been flushed from cache by intervening memory references. Thus the nominal goal in maximizing cache utilization is to reduce or eliminate capacity misses. We do not address the issue of *conflict* misses: given that a cache is associative and allowing a small percentage of the cache to remain apparently free when performing cache blocking, their impact is has proven unimportant. For architectures where their impact is significant various solutions exist [10].

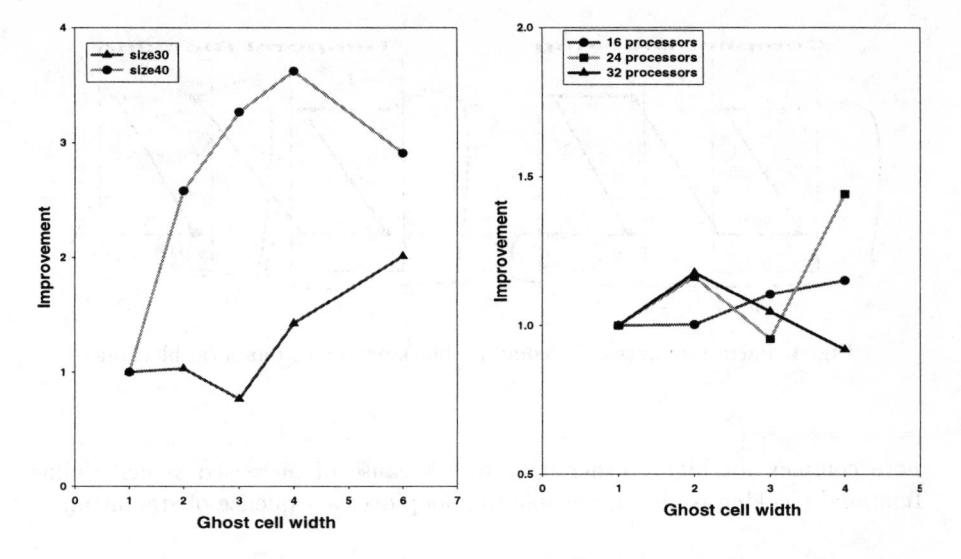

Fig. 2. Message aggregation: improvement as a function of problem size and ghost cell width

Fig. 3. Message aggregation: improvement as a function of number of processors and ghost cell width

To give an example of the relative speeds of the various levels of the memory hierarchy, on the Origin 2000—the machine for which we present performance data—the cost of accessing L1 cache is one clock cycle; 10 clock cycles for L2, 80 clock cycles for main memory, and for non-local memory 120 clock cycles plus network and cache-coherency overhead.

A problem with loops that multiply traverse an array (as in the given code fragment) is that when the array is larger than the cache, the data cycles repeately through the cache. This is common in numerical applications, and stencil operations in particular. *Cache blocking* seeks to increase temporal locality by re-ordering references to array elements so that small blocks that fit into cache undergo multiple traversals without intervening references to other parts of the array.

We distinguish two kinds of cache blocking: blocking done by a compiler (also called *tiling*) which we will refer to as *compiler blocking*, and our more effective technique, which we call *temporal blocking*, depicted in Fig. 3. In the case of e.g. stencil operations, a compiler won't do the kinds of optimizations we propose because of the dependence between outer iterations. A compiler may still perform blocking, but to a lesser effect. For both, the context in which the transformation may be applied is in sweeping over an array, typically in a simple regular pattern of access visiting each element using a stencil operator. Such operations are a common part of numerical applications, including more sophisticated numerical algorithms (e.g. multigrid methods). What we describe is independent of any particular stencil operator, though the technique becomes

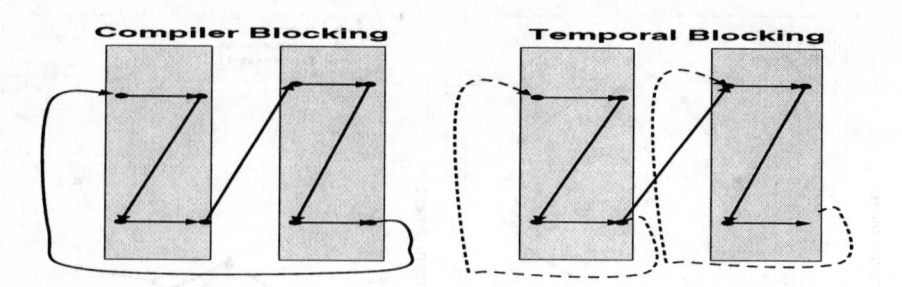

Fig. 4. Pattern of access for compiler blocking versus temporal blocking

more complex for higher-order operators because of increased stencil radius. Temporal blocking is also applicable to a loop over a sequence of statements.

5   The Temporal Blocking Algorithm

The basic idea behind the algorithm is that of applying a stencil operator to an array, *in place*, generalized to multiple applications of the stencil (iterations over the array) in such a way that only one traversal is required in each of one or more dimensions.

Consider first a stencil operator $f(x,y,z)$ applied to a 1D array $A[0..N]$. Ignoring treatment at the ends, the body of the loop, for loop index variable i, is

```
t = A[i];
A[i] = f( u, A[i], A[i+1] );
u = t;
```

Here t and u are the temporaries that serve the role of the array of initial data (or the previous iteration's values) for an algorithm that does not work in place.

Next we generalize to n iterations. For three iterations the code is

```
for (j=2; j!=-1; j--) {
  t[j] = A[i+j];
  A[i+j] = f( u[j], A[i+j], A[i+j+1] );
  u[j] = t[j];
}
```

First we observe that the 'window' into the array—here $A[i..i+3]$, may be as small as the stencil radius plus the number of iterations, as is the case here. Second, at the cost of slightly greater algorithmic complexity, so saving space but with small cost in time, only one temporary array is required, of length one greater than the minimum size of the window, rather than twice the minimum size of the window.

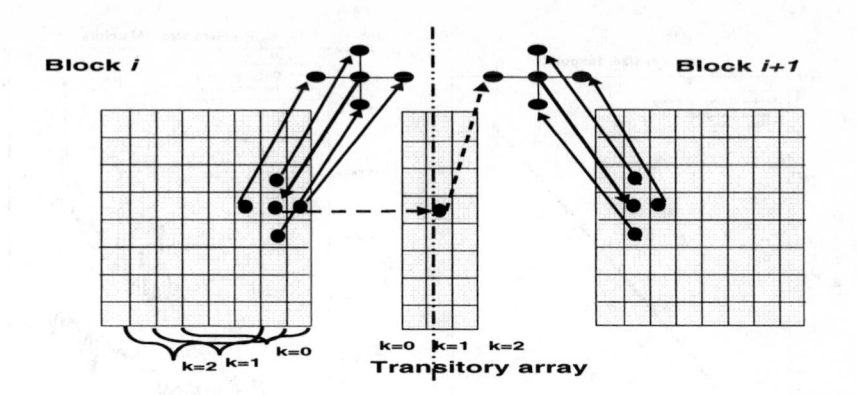

Fig. 5. Stencil operation and temporary storage for 1D decomposition of a 2D problem

It is this window into the array that we wish to be cache-resident. It may be any size greater than the minimum (the temporary storage requirements do not change); for our performance experiments the various arrays are sized so that they nearly fill the L1 cache.

Another observation is that given an n-dimensional problem with an m-dimensional decomposition, this technique may be applied with respect to any subset of the m dimensions of the decomposition—the more common and more simply coded multiple-traversal and/or old-new approximations approach applied to the remaining dimensions. The goal is to make all applications of the stencil operator to any given data element for a single cache miss (the compulsory one) for that element, which indicates that for larger problem size (relative to cache size) the technique must be applied with respect to a larger number of dimensions. Figure 5 depicts the stencil operation and temporary storage for a 1D decomposition of a 2D problem.

6 Performance Analysis

It is possible to predict the number of cache misses generated by the Jacobi relaxation code. In the case of compiler blocking the first sweep through the array should generate a number of cache misses equal to the number of elements accessed; these are compulsory. Each subsequent sweep will generate the same number of capacity misses. For temporal blocking only the compulsory misses should be generated. Experimental results shown in Fig. 6 bear this out. The data were collected on a MIPS R10000-based Silicon Graphics Inc. Origin 2000 with 32K of primary data cache and 4M of secondary unified cache, using on-processor hardware performance monitor counters; the programs were compiled at optimization level 3 using the MIPSpro C++ compiler. Fig. 7 contrasts the performance of compiler blocking and temporal blocking in terms of CPU cycles. The block size as well as the number of Jacobi iterations varies along the x-axis.

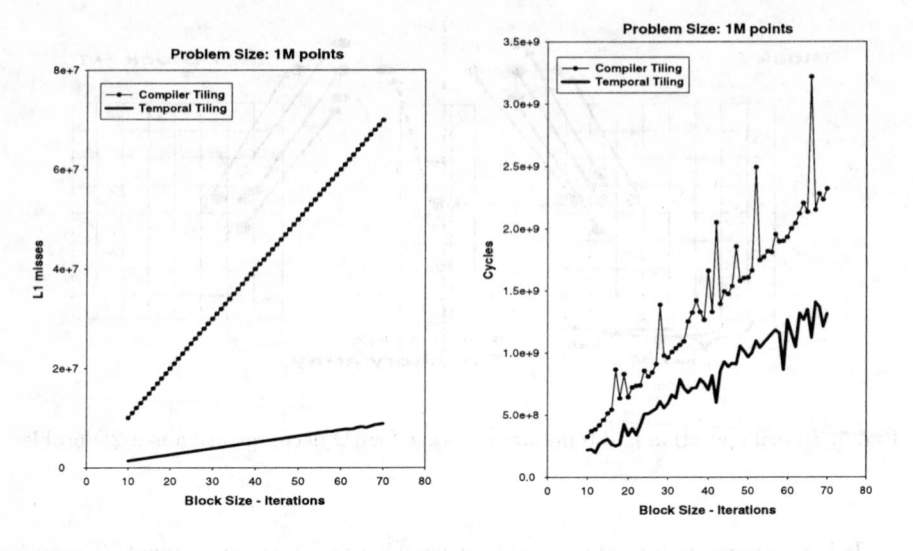

Fig. 6. L1 misses as a function of block size/number of iterations

Fig. 7. CPU cycles as a function of block size/number of iterations

Figure 7 shows that the temporal blocking version is twice as fast as the compiler blocking version until the block size exceeds the size of primary cache, beyond which temporal blocking and compiler blocking generate a similar number of cache misses. As expected, temporal blocking yields an improvement in performance linear in the number of iterations so long as the various data associated with a particular block fit in cache. Figure 8 shows that there is an ideal block size (relative to cache size)—in terms of CPU cycles, blocks smaller than ideal suffer from the constant overhead associated with the sweep of a single block; blocks larger than ideal generate capacity misses. (The spikes are attributable to anomalies of the hardware counters that do not capture results absolutely free of errors.)

The presence of multiple cache levels requires no special consideration: other experiments show optimization with respect to L1 cache is all that is required, and optimization with respect to L2 only is not nearly as beneficial.

For problem sizes exceeding the size of L2 (usually the case for meaningful problems), a straightforward implementation gives rise to a number of cache misses proportional to the number of iterations; with our transformation the number of misses is effectively constant for a moderate numbers of iterations.

In all the experiments for which results are given the block size is the same as the number of iterations. In this implementation of temporal blocking the number of possible iterations is limited by the block size. However, in most cases the number of iterations is dictated by the numerical algorithm. The choice then becomes that of the best block size given a fixed number of iterations. For a Jacobi problem a good rule of thumb is to have the number of elements in the transitory array be a small fraction of the number of elements that could fit in a block that fits in primary cache.

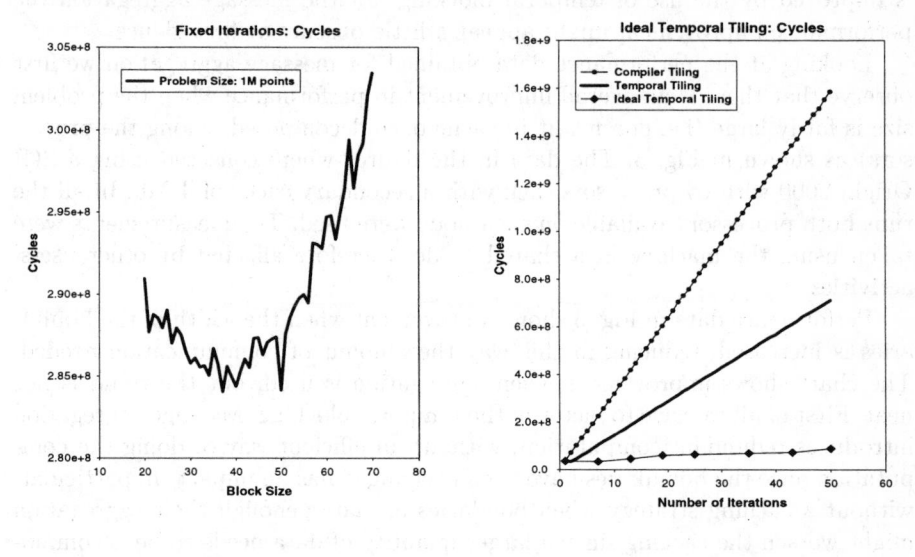

Fig. 8. CPU cycles as a function of block size for fixed number of iterations

Fig. 9. Relative predicted performance assuming ideal cache behaviour

The figures show, and Fig. 9 makes clear, that achieved performance improvement is not as good as predicted—performance improves with the number of iterations, but never exceeds a factor of two. The figure shows that the achieved miss behavior is not relatively constant, but depends on the number of iterations. The test code currently ignores the fact that permanent residency in cache for the transitory array cannot be guaranteed just by ensuring that there is always enough cache space for all the subsets of the arrays. Different subsets of the other arrays can map to the same locations in cache as does the transitory array, resulting in a significant increase in the number of conflict misses; this is a recognized problem with cache behavior; a solution is suggested in [10]. Various approaches are being evaluated under the assumption that there is still room for improvement before reaching some physical architecture-dependent limitations.

6.1 Message Aggregation

The message aggregation optimization is of relevance in the parallel context. It is important to point out that it would not be possible to enable the temporal blocking without the message aggregation. Thus, from a performance point of view the first expection is to not lose performance when using message aggregation. At this stage of investigation we believe that temporal blocking as a stand alone optimization has a greater impact on the overall performance than just message aggregation. The reasons in support of this are affected also by implementation issues. Currently, the two optimizations are implemented and tested separately. While it is shown clearly by the result collected how performance is improved by the use of temporal blocking, for the message aggregation the performance improvement might appear a little obscure at first glance.

Looking at the performance data obtained for message aggregation we first observe that there is an overall improvement in performance when the problem size is fairly large (i.e. doesn't fit in cache once decomposed among the processors) as shown in Fig. 3. The data in the figures where collected using a SGI Origin 2000 with 64 processors each with a secondary cache of 4 Mb. In all the runs both processors available on each node were used. The measurements were taken using the machine in a shared mode, therefore affected by other users' activities.

Performance data in Fig. 3 show improvement when the width of the boundaries is increased, reducing in this way the amount of communication needed. The chart shows improvement when aggregation is used, but the trend is not neat. First of all we have to factor in the temporal blocking. Message aggregation introduces redundant computation, without an efficient way of doing the computation once the boundaries have been exchanged has an impact. In particular, without a caching strategy when boudaries are large enough their aggregation might worsen the caching since a larger quantity of data needs to be accommodated in a local processor's cache. This translates into a reduction of the overall improvement. The purpose of the results presented in Figure 2 supports this intuition. A small problem size that always fits in cache has been used varying the size of the boundaries, but just on two processors (similar results can be obtained for a larger number of processor with the only difference being the problem size). The performance data show larger improvements with a more clear pattern. With this data the potential of the message aggregation optimization is more clear. When computation and communication are of roughly the same order this technique enable reduces the communication overhead which translate in a bigger potential for hiding the communication costs.

In this work we have only presented data obtained using an SGI Origin 2000 system. Origin systems have a very good network topology as well as very good latency and bandwidth. In particular, for a neighbor-to-neighbor communication pattern Origin systems perform particularly well even without message aggregation. In the test codes used the communication costs are negligible when the processors involved in the exchange are in the same node. It is clear that when data need to be exchanged with more processors that the optimization proposed will have a greater impact. Preliminary results obtained on a networked collection of Sun workstations support this claim.

7   Automating the Transformation

An optimizing transformation is generally only of academic interest if it is not deployed and used. In the context of array classes, it does not appear possible to provide this sort of optimization within the library itself because the applicability of the optimization is context dependent—the library can't know how its objects are being used. Two mechanisms for automating such optimizations are being actively developed: the use of *expression templates* (e.g. in POOMA), which seems too limited; and a source-to-source tranformational system (a preprocessor), which we are currently developing.

The ROSE preprocessor is a mechanism for (C++) source-to-source transformation, specifically targetted at optimizing the use of statements manipulated array class objects. It is based on the Sage II C++ source code restructuring tools and provides a distinct (and optional) step in the compilation process. It recognizes the use of the A++/P++ array class objects, and is 'hard-wired' with (later parameterized by) the A++/P++ array class semantics, so obviating the need for difficult or impossible program analysis. It is also parameterized by platform properties such as cache size. There is in principle no limit (within the bounds of computability) on the types of transformations that can be performed using this mechanism.

8 Conclusions

Previous work has focused on the optimization of the array class libraries themselves, and the use of techniques such as expression templates to provide better performance than the usual overloaded binary operators. We posit that such approaches are inadequate, that desirable optimizations exist that cannot be implemented by such methods, and that such approaches cannot reasonably be expected to be implemented by a compiler. One such optimization for cache architectures has been detailed and demonstrated.

A significant part of the utility of this transformation is in its use to optimize array class statements (a particularly simple syntax for the user which hides the parallelism, distribution, and communication issues) and in the delivery of the transformation through the use of a preprocessing mechanism.

The specific transformation we introduce addresses the use of array statements or collections of array statements within loop structures, thus it is really a family of transformations. For simplicity, only the case of a single array in a single loop has been described. Specifically, we evaluate the case of a stencil operation in a for loop. We examine the performance using the C++ compiler, but generate only C code in the transformation. We demonstrate that the temporal blocking transform is two times faster than the standard implementation.

The temporal blocking transformation is language independent, although we provide no mechanism to automate the transformation outside of the Overture object-oriented framework. The general approach could equally well be used with FORTRAN 90 array syntax.

Finally, the use of object-oriented frameworks is a powerful tool, but limited in use by the performance being less than that of FORTRAN 77; we expect that work such as this to change this situation, such that in the future one will use such object-oriented frameworks because they represent *both* a higher-level, simpler, and more productive way to develop large-scale applications *and* a higher performance development strategy. We expect higher performance because the representation of the application using the higher level abstractions permits the use of new tools (such as the ROSE optimizing preprocessor) that can introduce more sophisticated transformation (because of their more restricted semantics) than compilers could introduce (because of the broader semantics that the complete language represents).

References

[1] David Brown, Geoff Chesshire, William Henshaw, and Dan Quinlan. Overture: An object-oriented software system for solving partial differential equations in serial and parallel environments. In *Proceedings of the SIAM Parallel Conference*, Minneapolis, MN, March 1997.

[2] J.V.W. Reynders et. al. POOMA: A framework for scientific simulations on parallel architectures. In *Parallel Programming using C++* by Gregory V. Wilson and Paul Lu, MIT Press, 1996.

[3] Federico Bassetti, Kei Davis, and Dan Quinlan. Toward Fortran 77 performance from object-oriented scientific frameworks. In *Proceedings of the High Performance Computing Conference (HPC'98)*, 1998.

[4] Todd Veldhuizen. Expression templates. In S.B. Lippmann, editor, *C++ Gems*. Prentice-Hall, 1996.

[5] Steven S. Muchnick. *Advanced Compiler Design and Implementation*. Morgan Kaufmann, 1997.

[6] Rebecca Parsons and Dan Quinlan. A++/P++ array classes for architecture independent finite difference computations. In *Proceedings of the Second Annual Object-Oriented Numerics Conference (OONSKI'94)*, April 1994.

[7] Bjarne Stroustrup. *The C++ Programming Language*. Addison Wesley, third edition edition, 1997.

[8] Roldan Pozo. Template numerical toolkit. http:// math.nist.gov/ tnt/.

[9] GNU scientific software library. http:// KachinaTech.com.

[10] Naraig Manjikian and Tarek Abdelrahman. Array data layout for the reduction of cache conflicts. In *International Conference on Parallel and Distributed Computing Systems (PDCS'98)*, 1995.

Merging Web-Based with Cluster-Based Computing

Luís Moura Silva, Paulo Martins, João Gabriel Silva

Universidade de Coimbra - POLO II
Coimbra - Portugal

Abstract. This paper presents a Java-based software infrastructure that allows the merging of Web-based metacomputing with cluster-based parallel computing. We briefly compare the two approaches and we describe the implementation of a software bridge that supports the execution of meta-applications: some processes of the application may run over the Internet as Java applets while other processes of the same application can execute on a dedicated cluster of machines running PVM or MPI. We present some performance results that show the effectiveness of our approach.

1. Introduction

One decade ago the execution of parallel problems was dominated by the use of supercomputers, vector computers, multiprocessor and shared-memory machines. In this past decade there has been an increasing trend towards the use of Network of Workstations (NOWs) [1]. Another concept that recently has become popular is Web-based metacomputing [2]. The idea is to use geographically distributed computers to solve large parallel problems with the communication done through the Internet. In practice, the idea behind Web-based parallel computing is just a new variation over NOW-based computing: the recycling of idle CPU cycles in the huge amount of machines connected to the Internet.

Both approaches have their domain of application: cluster-based computing is used to execute a parallel problem that is of interest of some institution or company. The parallel problem is typically of medium-scale and the network of computers usually belongs to the same administrative domain. Cluster-based computing is now well established and libraries like PVM [3] and MPI [4] have been widely used by the HPC community. On the other hand, Web-based computing is more suited for the execution of long-running applications that have a global interest, like solving some problems of cryptography, mathematics and computational science. It uses computing resources that belong to several administrative domains. Web-based computing is not widely used, though there are already some successful implementations of this concept. Examples include the Legion project [5], Globus [6], Charlotte [7], Javelin [8] among others.

This work was partially supported by the Portuguese Ministry of Science and Technology (MCT) under program PRAXIS XXI.

In this paper, we present a software infrastructure that allows the simultaneous exploitation of both approaches. The system was developed in Java and makes part of the JET project [9]. Originally, the JET system provided a model of execution based on Internet computing: the applications are executed by Java applets and are downloaded through a standard Web-browser by the user who wants to volunteer some CPU cycles of his machine to a global computation. Lately, we have included the necessary features to support the execution of applications in cluster of computers running well-established communication libraries like MPI and PVM. These two libraries have been enhanced with a Java interface and we have developed a module that glues both models of computation. This module is called *JET-Bridge* and will be described in this paper.

The rest of the paper is organized as follows, section 2 presents the functional architecture of the JET system. Section 3 describes in some detail the JET-Bridge. Section 4 presents some performance results. Section 5 concludes the paper.

2. A General Overview of the JET Project

JET is a software infrastructure that supports parallel processing of CPU-intensive problems that can be programmed in the Master/Worker paradigm. There is a Master process that is responsible for the decomposition of the problem into small and independent tasks. The tasks are distributed among the Worker processes that execute a quite simple cycle: receive a task, compute it and send the result back to the Master. The Master is responsible for gathering the partial results and to merge them into the problem solution. Since each task is independent. there is no need for communication between worker processes.

The Worker processes execute as Java applets inside a Web browser. The user that wants to volunteer his spare CPU cycles to a JET computation just need to access a Web page by using a Java-enabled browser. Then, she just has to click somewhere inside the page and one Worker Applet is downloaded to the client machine. This Applet will communicate with a JET Master that executes on the same remote machine where the Web page came from.

The communication between the worker applets and the JET Master is done through UDP sockets. This class of sockets provides higher scalability and consumes fewer resources than TCP sockets. The UDP protocol does not guarantee the delivery of messages but the communication layer of JET implements a reliable service that insures sequenced and error-free message delivery. The library keeps a time-out mechanism for every socket connection in order to detect the failure or a withdrawn of some worker applet.

The JET system provides some internal mechanisms to tolerate the high latency of the communication over the Internet. Those techniques are based on the prefetching of tasks by the remote machines and the asynchronous flush of output results back to the JET Master. There are some internal threads that perform the communication in a concurrent way with the normal execution of the application processes.

The number of machines that can join a JET computation is surely unpredictable but the system should be able to manage hundreds or thousands of clients. To assure a scalable execution we depart from the single-server approach and the forthcoming version of JET relies in a hierarchical structure of servers, as represented in Figure 1.

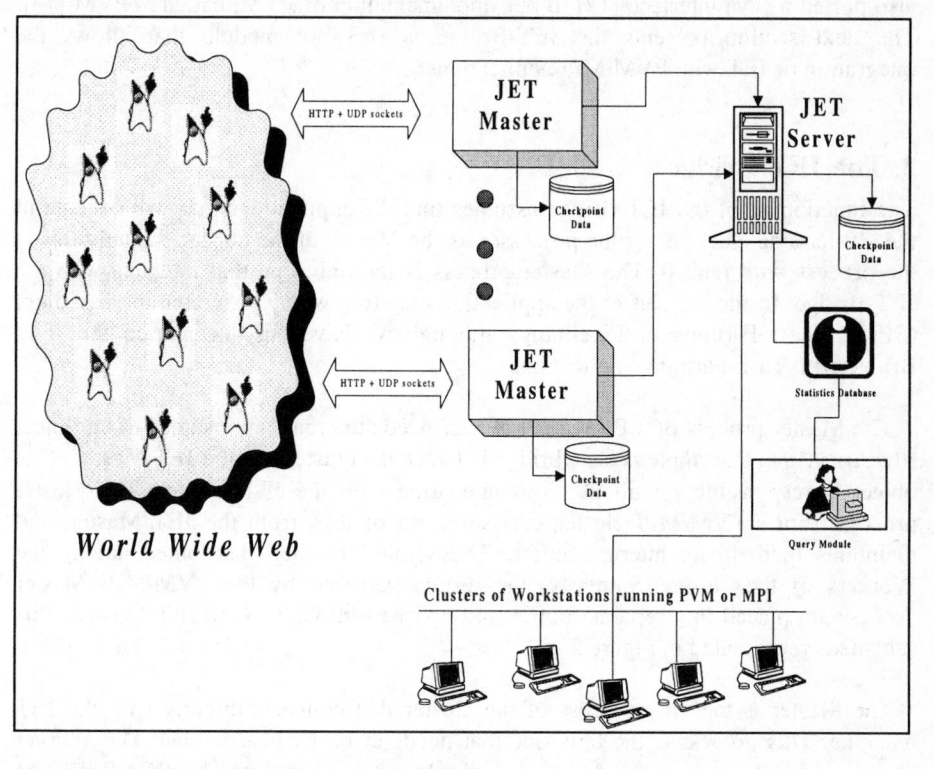

Fig. 1. The Structure of the JET virtual machine.

This scalable structure relies in multiple JET Masters: every Master will be responsible for a sub-set of worker machines dividing the load more evenly and increasing the reliability of the system. Every JET Master communicates with a centralized JET Server, which maintains the global status of the application and a database with all the interesting statistics.

The JET system includes some fault-tolerance mechanisms. Task reconfiguration is used to tolerate the loss of worker applet. The resiliency of the Master processes is achieved through the use of checkpointing and logging techniques. The checkpointing mechanism is necessary to assure the continuity of the application when there is a failure or a preventive shutdown of a JET Master or the main Server. The critical state of the application is saved periodically in stable storage in some portable format that allows its resumption later on in the same or in a different machine.

JET computations will not be restricted to the Web. It is possible to use some other existing high-performance computing resources, like cluster of workstations or parallel machines. The basic idea is to allow existing clusters of machines running PVM or MPI to inter-operate with a JET computation. To achieve this interoperability we have implemented a Java interface to the Windows version of MPI that was developed by our research group [10]. The Java binding is described in [11]. We have also ported a Java interface [12] to our implementation of PVM, called WPVM [13]. The next section presents the JET-Bridge, a software module that allows the integration of JET with PVM/MPI applications.

3. The JET-Bridge

The functioning of the JET-Bridge assumes that the applications that will execute in the cluster side elect one of the processes as the Master of the cluster. Usually this is the process with rank 0. The Master process is the only one that interacts with the JET-Bridge. Inside the cluster the application may follow any programming paradigm (SPMD, Task-Farming or Pipelining) although we have only been used the JET-Bridge with Task-Farming applications.

The Master process of a PVM/MPI cluster needs to create an instance of an object (JetBridge) that implements a bridge between the cluster and the JET Master. This object is responsible for all the communication with the JET Master. The Master process from a PVM/MPI cluster gets some set of jobs from the JET Master, and maintains them in an internal buffer. These jobs are then distributed among the Workers of the cluster. Similarly, the results gathered by the PVM/MPI Master process are placed in a separate buffer and will be sent later to the JET Master. This scheme is represented in Figure 2.

The Master is the only process of the cluster that connects directly with the JET machine. This process is the only one that needs to be written in Java. The Worker processes can be implemented in any of the languages supported by WMPI/WPVM libraries (i.e. C, Fortran, and Java) and all the heterogeneity is solved by using the Java bindings [11].

When the user creates an instance of the object JetBridge two new threads are created: the Sender thread and the Receiver thread. These threads are responsible for all the communication with the JET Master.

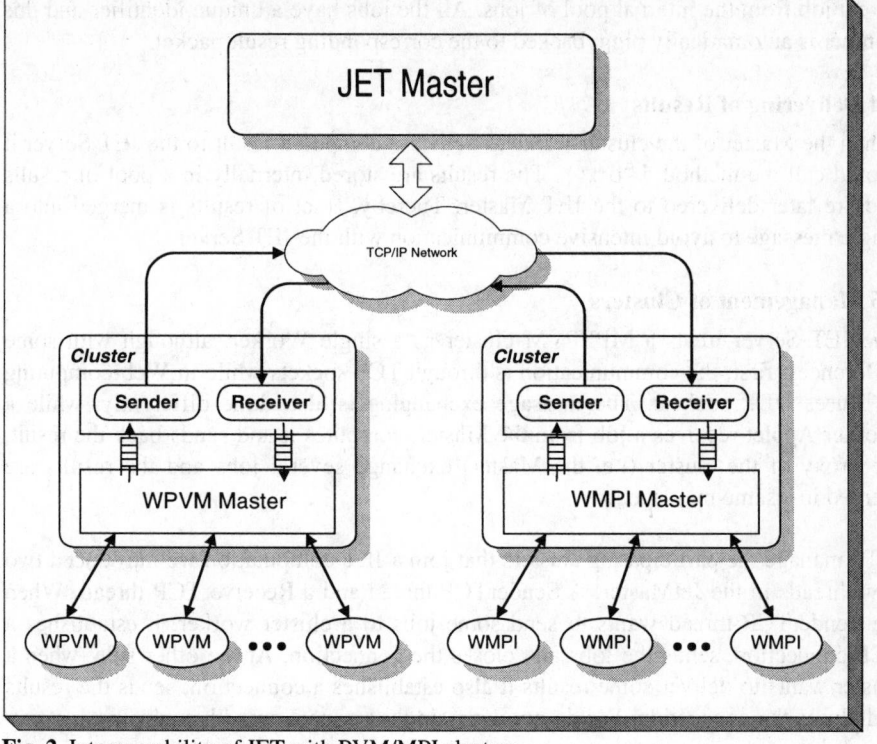

Fig. 2. Interoperability of JET with PVM/MPI clusters.

3.1 Registering to the JET Machine

To connect to the JET machine the WPVM or WMPI application should call the method jStart(). This method performs the registry with the JET machine. The JET Server creates a service entry for that cluster and assigns an identification key that is sent back to the WPVM or WMPI Master process (rank 0). After this registration process the cluster is willing to participate in a JET computation.

3.2 Obtaining Jobs to Compute

To mask the latency of the network and exploit some execution autonomy the JET-Bridge performs some job prefetching from the JET Server. A set of jobs is grabbed from the server and they are placed in an internal pool of jobs residing in the machine where the WPVM/WMPI Master process is running. Later on, these jobs are distributed among the Worker processes of the cluster. The number of jobs that are prefetched each time depends on the size of the cluster. When the JET-Bridge detects that the number of jobs to compute is less than the number of Workers it sends another request to the JET Server asking for a new bunch of jobs.

3.3 Distribution of Jobs

The Master of the cluster is responsible for the distribution of the jobs among the rest of the application processes. The Master should make a call to the method jGet() to take a job from the internal pool of jobs. All the jobs have a unique identifier and this number is automatically piggybacked to the corresponding result packet.

3.4 Delivering of Results

When the Master of the cluster wants to deliver a computed result to the JET Server it should call the method jPut(). The results are stored internally in a pool of results and are later delivered to the JET Master. Thereby, a set of results is merged into a single message to avoid intensive communication with the JET Server.

3.5 Management of Clusters

The JET Server treats a MPI/PVM cluster as a single Worker, although with some differences: first, the communication is through TCP sockets while in Web-computing JET uses UDP sockets. The message exchanging is also done differently: while a Worker Applet receives a job from the Master, computes it and sends back the result, the proxy of the cluster (i.e. the Master) exchange several jobs and the results are merged in a same message.

To manage the participating clusters that join a JET computation we introduced two new threads in the JetMaster. A SenderTCP thread and a ReceiverTCP thread. When the SenderTCP thread wants to send some jobs to a cluster worker, it establishes a TCP connection, sends the jobs and closes the connection. At the other side, when a cluster wants to deliver some results it also establishes a connection, sends the results and closes the connection. We do not maintain the connection with each cluster proxy to avoid the exhaustion of system resources.

The JET-Bridge infrastructure is totally independent from the underlying application. In the future, if we want to port other well-known parallel libraries to Java, it will be easy to integrate them with the JET platform.

4. Performance Results

In this section we present some results of an experimental study that show the effectiveness of the JET-Bridge together with the Java bindings that we have implemented. All the measurements were taken with the NQueens benchmark with 14 queens in a cluster of Pentiums 200MHz running Windows NT 4.0, which are connected through a non-dedicated 10 Mbit/sec Ethernet.

Figure 3 presents several different combinations of heterogeneous configurations of this benchmark using WMPI and the JET platform. JWMPI corresponds to the case where the application was written in Java and used the WMPI library. In CWMPI the application written in C, while JWMPI(Native) represents a hybrid case: the interface to the WMPI was written in Java but the real computation was done by a native method written in C.

The first three experiments present the results of computations using homogeneous clusters of 8 processes. As we can see from the Figure the two computations that used native code (i.e. C) executed in about half a time of the pure Java version. This is due to the difference in performance presented by the languages. Until the Java compiler technology reaches maturity, the use of native code in Java programs is possible a way to improve performance. Providing access to standard libraries, often required in scientific programming, seems imperative in order to allow the reuse of existing code that was developed with MPI and PVM.

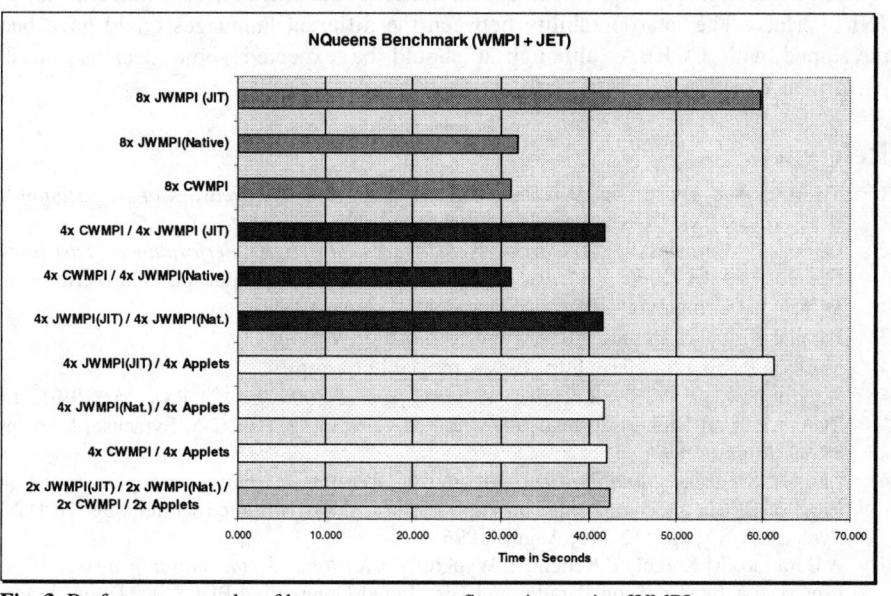

Fig. 3. Performance results of heterogeneous configurations using WMPI.

In our implementation of the NQueens benchmark the jobs are distributed on demand, allowing the faster workers to compute more jobs than the slower ones. So, the best performance is obtained in all the computations that include processes entirely written in C or those hybrid processes (Java+C) that use the native version of the kernel.

More than the absolute results, this experiment has proved the importance of the JET-Bridge and the Java bindings, which allow us to exploit the potential of a really heterogeneous computation. Where some processes were executing as Java Applets, others may execute in Java and use the WMPI library. They can also interoperate in the same application with other processes written in C or that use a hybrid approach: Java and native code.

5. Conclusions

The system presented in this paper provides an integrated solution to unleash the potential of diverse computational resources, different languages and different execution models. With a software module like the JET-Bride it would be possible to execute some applications where part of the tasks are executed over the Internet while other tasks are computed on a dedicated PVM or MPI cluster-based platform.

Moreover, some of the processes of the application are executed as Java Applets, other processes are written in C; others may be written in Java and use native or full Java code. The cluster processes can also chose between the MPI and the PVM API. There could be some meta-applications that would exploit this potential of interoperability.

The JET Bridge was implemented with socket communication and made use of the JNI interface. The interoperability between the different languages could have been developed with CORBA, although it should be expected some decrease in the performance.

References

1. T.E.Anderson, D.E.Culler, D.A.Patterson. *"A Case for NOW (Network of Workstations)"*, IEEE Micro, Vol. 15. No.1, pp. 54-64, February 1995
2. G.Fox, W.Furmanski. *"Towards Web/Java-based High Performance Distributed Computing - An Evolving Virtual Machine"*, Proc. 5th Symposium on High-Performance Distributed Computing, HPDC-5, Syracuse, NY, August 1996
3. Parallel Virtual Machine, http://www.epm.ornl.gov/pvm/
4. Message Passing Interface, http://www.mcs.anl.gov/mpi/
5. A.Grimshaw, W.Wulf. *"Legion - A view from 50,000 Feet"*, Proc. 5th IEEE Int. Symposium on High-Performance Distributed Computing, HPDC-5, Syracuse, USA, pp. 89-99, August 1996
6. I.Foster, S.Tuecke. *"Enabling Technologies for Web-Based Ubiquitous Supercomputing"*, Proc. 5th IEEE int. Symposium on High-Performance Distributed Computing, HPDC-5, Syracuse, USA, pp. 112-119, August 1996
7. A.Baratloo, M.Karaul, Z.Kedem, P.Wyckoff. *"Charlotte: Metacomputing on the Web"*, Proc. ISCA Int. Conf. on Parallel and Distributed Computing, PDCS'96, Dijon, France, pp.181-188, Sept. 1996
8. P.Cappelo, B.Christiansen, M.F.Ionescu, M.Neary, K.Schauser, D.Wu. *"Javelin: Internet-based Parallel Computing using Java"*, ACM 1997 Workshop on Java for Science and Engineering Computation, Las Vegas, USA, June 1997
9. L.M.Silva, H.Pedroso, J.G.Silva. *"The Design of JET: A Java Library for Embarrassingly Parallel Applications"*, WOTUG'20 - Parallel Programming and Java Conference, Twente, Netherlands, 1997
10. Windows Message Passing Interface, http://dsg.dei.uc.pt/wmpi/
11. P.Martins, L.M.Silva, J.G.Silva. *"A Java Interface for WMPI"*, to appear in EuroPVM/MPI98, Liverpool, UK, September 1998.
12. JavaPVM Homepage, http://homer.isye.gatech.edu/chmsr/JavaPVM/
13. A.Alves, L.M.Silva, J.Carreira, J.G.Silva, *"WPVM: Parallel Computing for the People"*, Proc. of HPCN'95, High Performance Computing and Networking Europe, May 1995, Milano, Italy, Lecture Notes in Computer Science 918, pp. 582-587

Dynamic Reconfiguration and Virtual Machine Management in the Harness Metacomputing System

Mauro Migliardi[1], Jack Dongarra[2,3], Al Geist[2], and Vaidy Sunderam[1]

[1] Emory University, Atlanta, GA, USA
[2] Oak Ridge Natonal Laboratory, Oak Ridge, TN, USA
[3] University of Tennessee, Knoxville, TN, USA

Abstract. Metacomputing frameworks have received renewed attention of late, fueled both by advances in hardware and networking, and by novel concepts such as computational grids. However these frameworks are often inflexible, and force the application into a fixed environment rather than trying to adapt to the application's needs. Harness is an experimental metacomputing system based upon the principle of dynamic reconfigurability not only in terms of the computers and networks that comprise the virtual machine, but also in the capabilities of the VM itself. These characteristics may be modified under user control via a "plug-in" mechanism that is the central feature of the system. In this paper we describe how the design of the Harness system allows the dynamic configuration and reconfiguration of virtual machines, including naming and addressing methods, as well as plug-in location, loading, validation, and synchronization methods.

1 Introduction

Harness is an experimental metacomputing system based upon the principle of dynamically reconfigurable networked computing frameworks. Harness supports reconfiguration not only in terms of the computers and networks that comprise the virtual machine, but also in the capabilities of the VM itself. These characteristics may be modified under user control via a "plug-in" mechanism that is the central feature of the system. The motivation for a plugin-based approach to reconfigurable virtual machines is derived from two observations. First, distributed and cluster computing technologies change often in response to new machine capabilities, interconnection network types, protocols, and application requirements. For example, the availability of Myrinet [[1]] interfaces and Illinois Fast Messages has recently led to new models for closely coupled Network Of Workstations (NOW) computing systems. Similarly, multicast protocols and better algorithms for video and audio codecs have led to a number of projects that focus on tele-presence over distributed systems. In these instances, the underlying middleware either needs to be changed or re-constructed, thereby increasing the effort level involved and hampering interoperability. A virtual machine model intrinsically incorporating reconfiguration capabilities will address these issues in an effective manner. The second reason for investigating the plug-in model is to attempt to provide a virtual machine environment that can dynamically adapt to meet an application's needs, rather than forcing the application to fit into a fixed environment. Long-lived simulations evolve through several phases: data input, problem setup, calculation, and analysis or visualization of results. In traditional, statically configured metacomputers, resources needed during one phase are often underutilized in other phases. By allowing applications to dynamically reconfigure the system, the overall utilization of the computing infrastructure can be enhanced.

The overall goals of the Harness project are to investigate and develop three key capabilities within the framework of a heterogeneous computing environment:

- Techniques and methods for creating an environment where multiple distributed virtual machines can collaborate, merge or split. This will extend the current network and cluster computing model to include multiple distributed virtual machines with multiple users, thereby enabling standalone as well as collaborative metacomputing.
- Specification and design of plug-in interfaces to allow dynamic extensions to a distributed virtual machine. This aspect involves the development of a generalized plug-in paradigm for distributed virtual machines that allows users or applications to dynamically customize, adapt, and extend the distributed computing environment's features to match their needs.
- Methodologies for distinct parallel applications to discover each other, dynamically attach, collaborate, and cleanly detach. We envision that this capability will be enabled by the creation of a framework that will integrate discovery services with an API that defines attachment and detachment protocols between heterogeneous, distributed applications.

In the preliminary stage of the Harness project, we have focused upon the dynamic configuration and reconfiguration of virtual machines, including naming and addressing schemes, as well as plugin location, loading, validation, and synchronization methods. Our design choices, as well as the analysis and justifications thereof, and preliminary experiences, are reported in this paper.

2 Architectural Overview of Harness

The architecture of the Harness system is designed to maximize expandability and openness. In order to accommodate these requirements, the system design focuses on two major aspects: the management of the status of a Virtual Machine that is composed of a dynamically changeable set of hosts; the capability of expanding the set of services delivered to users by means of plugging into the system new, possibly user defined, modules on-demand without compromising the consistency of the programming environment.

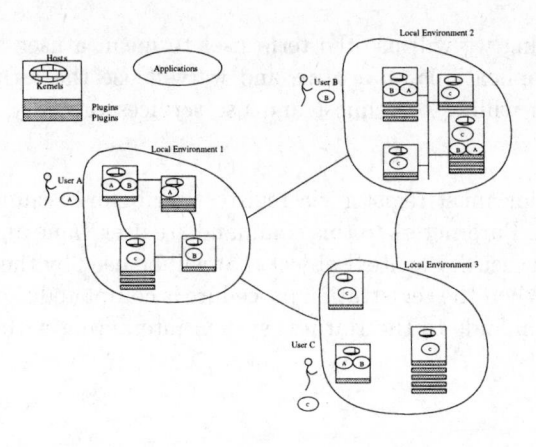

Fig. 1. A Harness virtual machine

2.1 Virtual Machine Startup and Harness System Requirements

The Harness system allows the definition and establishment of one or more Virtual Machines (VMs). A Harness VM (see Fig. 1) is a distributed system composed of a VM status server and a set of kernels running on hosts and delivering services to users.

The current prototype of the Harness system implements both the kernel and the VM status server as pure Java programs. We have used the multithreading capability of the Java Virtual Machine to exploit the intrinsic parallelism of the different tasks the programs have to perform, and we have built the system as a package of several Java classes. Thus, in order to be able to use the Harness system a host should be capable of running Java programs (i.e. must be JVM equipped). The different components of the Harness system communicates through reliable unicast channels and unreliable multicast channels. In the current prototype these communication commodities are implemented using the java.net package.

In order to use the Harness system, applications should link to the Harness core library. The basic Harness distribution will include core library versions for C, C++ and Java programs but in the following description we show only Java prototypes.

This library provides access to the only hardcoded service access point of the Harness system, namely the core function

```
Object H_command(String VMSymbolicName, String[] theCommand).
```

The first argument to this function is a string specifying the symbolic name of the virtual machine the application wants to interact with. The second argument is the actual command and its parameters. The command might be one of the User Kernel Interface commands as defined later in the paper or the registerUser command. The return value of the core function depends on the command issued.

In the following we will use the term user to mean a user that runs one or more Harness applications on a host, and we will use the term application to mean a program willing to request and use services provided by the Harness system.

Any application must register via registerUser before issuing any command to a Harness VM. Parameters to this command are userName and userPassword; this call will set a security context object that will be used by the system to check user privileges. When the registration procedure is completed the application can start issuing commands to the Harness system interacting with a local Harness kernel.

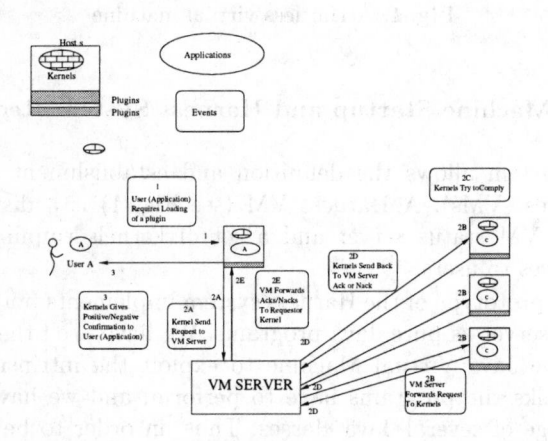

Fig. 2. Event sequence for a distributed plug-in loading

A Harness kernel is the interface between any application running on a host and the Harness system. Each host willing to participate in a Harness VM runs one kernel for each VM. The kernel is bootstrapped by the core library during the user registration procedure. A Harness kernel delivers services to user programs and cooperates with other kernels and the VM status server to manage the VM. The status server acts as a repository of a centralized copy of the VM status and as a dispatcher of the events that the kernel entities want to publish to the system (see Fig. 2). Each VM has only one status server entity in the sense that all the other entities (kernels) see it as a single monolithic entity with a single access point. Harness VM's use a built-in communication subsystem to distribute system events to the participating active entities. Applications based on message passing may use this substrate or may provide their own communications fabric in the form of a Harness plug-in. In the prototype, native communications use TCP and UDP/IP-multicast.

2.2 Virtual Machine Management: Dynamic Evolution of a Harness VM

In our early prototype of Harness, the scheme we have developed for maintaining the status of a Harness VM is described below. The status of each VM is composed of the following information:

- Membership: the set of participating kernels;
- Services: the set of services that, based on the plug-in modules currently loaded, the VM is able to perform both as a whole and on a per-kernel basis;
- Baseline: the services that new kernels needs to be able to deliver to join the VM and the semantics of these services;

It is important to notice that the VM status is kept completely separated from the internal status of any user application in order to prevent its consistency protocol from constraining users' applications requirements.

To prevent the status server from being a single point of failure, each VM in the Harness system keeps two copies of its status: one is centralized in the status server and the second collectively maintained among the kernels. This mechanism allows reconstruction of the status of each crashed kernel from the central copy and, in case of status server crash, reconstructing the central copy from the distributed status information held among the kernels.

Each Harness VM is identified by a VM symbolic name. Each VM symbolic name is mapped onto a multicast address by a hashing function. A kernel trying to join a VM multicasts a "join" message on the multicast address obtained by applying the hashing function to the VM symbolic name. The VM server responds by connecting to the inquiring kernel via a reliable unicast channel, checking the kernel baseline and sending back either an acceptance message or a rejection message. All further exchanges take place on the reliable unicast channel. To leave a VM a kernel sends a "leave" message to the VM server. The VM server publishes the event to all the remaining kernels and updates the VM status. Every service that each kernel supports is published by the VM status server to every other kernel in the VM. This mechanism allows each kernel in a Harness VM to define the set of services it is interested in and to keep a selective up-to-date picture of the status of the whole VM. Periodic "I'm alive" messages are used to maintain VM status information; when the server detects a crash, it publishes the event to every other kernel. If and when the kernel rejoins, the VM server gives it the old copy of the status and wait for a new, potentially different, status structure from the rejoined kernel. The new status is checked for compatibility with current VM requirements. A similar procedure is used to detect failure of the VM server and to regenerate a replacement server.

2.3 Services: The User Interface of Harness Kernels

The fundamental service delivered by a Harness kernel is the capability to manipulate the set of services the system is able to perform. The user interface of Harness kernels accepts commands with the following general syntax:

```
<command> <locator> <targets>
    <Quality of Service> [additional parameters]
```

The command field can contain one of the following values:

- Load to install a plug-in into the system;
- Run to run a thread to execute plug-in code;
- Unload to remove an unused plug-in from the system;
- Stop to terminate the execution of a thread

Services delivered by plug-ins may be shared according to permission attributes set on a per plug-in basis. Users may remove only services not in the core category. A core service is one that is mandatory for a kernel to interact with the rest of the VM. With the stop and unload commands a user can reclaim resources from a service that is no longer needed, but the nature of core services prevents any user from downgrading a kernel to an inoperable state. However, although it is not possible to change core services at run time, they do not represent points of obsolescence in the Harness system. In fact they are implemented as hidden plug-in modules that are loaded into the kernel at bootstrap time and thus easily upgraded. The core services of the Harness system form the baseline and must be provided by each kernel that wishes to join a VM. They are:

- The VM server crash recovery procedure;
- The plug-in loader/linker module;
- The core communication subsystem.

Commands must contain the unique locator of the plug-in to be manipulated. The lowest level Harness locator, the one actually accepted by the kernel, is a Uniform Resource Locator (URL). However any user may load at registration time a plug-in module that enhances the resource management capabilities of the kernel by allowing users to adopt Uniform Resource Names (URNs), instead of URLs, as locators. The version of this plugin provided with the basic Harness distribution allows:

- Checking for the availability of the plug-in module on multiple local and remote repositories (e.g. a user may simply wish to load the "SparseMatrix-Solver" plug-in without specifying the implementation code or its location);
- The resolution of any architecture requirement for impure-Java plug-ins.

However, the level of abstraction at which service negotiation and URN to URL translation will take place, and the actual protocol implementing this procedure, can be enhanced/changed by providing a new resource manager plug-in to kernels.

The target field of a command defines the set of kernels that are required to execute the command. Every non-local command is executed using a two phase commit protocol. Each command can be issued with one of the following Quality of Service(QoS): all-or-none and best-effort. A command submitted with a all-or-none QoS succeeds if and only if all of the kernels specified in the target field are able (and willing) to execute it. A command submitted with a best-effort QoS fails if and only if all the kernels specified in the target field are unable (unwilling) to execute it. Both the failure and the success return values include the list of kernel able (willing) to execute the command and the list of the unable (unwilling) ones.

2.4 Related Work

Metacomputing frameworks have been popular for nearly a decade, when the advent of high end workstations and ubiquitous networking in the late 80's enabled high performance concurrent computing in networked environments. PVM [2] was one of the earliest systems to formulate the metacomputing concept in concrete terms, and explore heterogeneous network computing. PVM however, is inflexible in many respects. For example, multiple DVM merging and splitting is not supported. Two different users cannot interact, cooperate, and share resources and programs within a live PVM machine. PVM uses internet protocols which may preclude the use of specialized network hardware. A "plug-in" paradigm would alleviate all these drawbacks while providing greatly expanded scope and substantial protection against both rigidity and obsolescence.

Legion [3] is a metacomputing system that began as an extension of the Mentat project. Legion can accommodate a heterogeneous mix of geographically distributed high-performance machines and workstations. Legion is an object oriented system where the focus is on providing transparent access to an enterprise-wide distributed computing framework.

The model of the Millennium system [4] being developed by Microsoft Research is similar to that of Legion's global virtual machine. Logically there is only one global Millennium system composed of distributed objects. However, at any given instance it may be partitioned into many pieces. Partitions may be caused by disconnected or weakly-connected operations. This could be considered similar to the Harness concept of dynamic joining and splitting of DVMs.

Globus [5] is a metacomputing infrastructure which is built upon the Nexus [6] communication framework. The Globus system is designed around the concept of a toolkit that consists of the pre-defined modules pertaining to communication, resource allocation, data, etc. Globus even aspires to eventually incorporate Legion as an optional module. This modularity of Globus remains at the metacomputing system level in the sense that modules affect the global composition of the metacomputing substrate.

The above projects envision a much wider-scale view of distributed resources and programming paradigms than Harness. Harness is not being proposed as a world-wide infrastructure, but more in the spirit of PVM, it is a small heterogeneous distributed computing environment that groups of collaborating scientists can use to get their science done. Harness is also seen as a research tool for exploring pluggability and dynamic adaptability within DVMs.

3 Conclusions and Future Work

In this paper we have described our early work on the plug-in mechanism and the dynamic Virtual Machine (VM) management mechanism of the Harness system, an experimental metacomputing system. These mechanisms allow the Harness system to achieve reconfigurability not only in terms of the computers and networks that comprise the VM, but also in the capabilities and the services provided by the VM itself, without compromising the coherency of the programming environment.

Early experience with small example programs show that the system is able to:

- Adapt to changing user needs by adding new services via the plug-in mechanism;
- Safely add or remove services to a distributed VM;
- Locate, validate and load locally or remotely stored plug-in modules;
- Cope with network and host failure with a limited overhead;
- Dynamically add and remove hosts to the VM via the dynamic VM management mechanism.

In a future stage of the Harness project we will test these feature on real world applications.

References

[1] N. Boden et al. MYRINET: a gigabit per second local area network. *IEEE-Micro*, Vol. 15, No. 1, February 1995.

[2] A. Geist, A. Beguelin, J. Dongarra, W. Jiang, B. Manchek, and V. Sunderam. *PVM: Parallel Virtual Machine a User's Guide and Tutorial for Networked Computing*, MIT Press, Cambridge, MA, 1994.

[3] A. Grimshaw, W. Wulf, J. French, A. Weaver, and P. Reynolds. Legion: the next logical step toward a nationwide virtual computer. Technical Report CS-94-21, University of Virginia, 1994.

[4] Microsoft Corporation. Operating systems directions for the next millenium. Position paper available at
http://www.research.microsoft.com/research/os/Millenium/mgoals.html.

[5] I. Foster and C. Kesselman. Globus: a metacomputing infrastructure toolkit. *Int. J. Supercomputing Applications*, May 1997.

[6] I. Foster, C. Kesselman, and S. Tuecke. The Nexus approach to integrated multithreading and communication. *J. Parallel and Distributed Computing*, 37:70-82, 1996.

JEM-DOOS: The Java*/RMI Based Distributed Objects Operating System of the JEM Project**

Serge Chaumette

LaBRI, Laboratoire Bordelais de Recherche en Informatique, U. Bordeaux I, France

Abstract. The Java technology[1] provides support to design and develop platforms to deal with heterogeneous networks. One of the goals of the *JEM* project, *Experimentation environMent for Java*, carried out at LaBRI is to design and develop such a platform. The JEM project[2] consists in: providing a distributed platform that makes using heterogeneous networks of computers easier; using this platform as a laboratory for experimentation purpose. It is based on Java, RMI[3] and CORBA[4]. In this paper, we present an overview of the conception and the implementation of the kernel of our platform. This kernel is called *JEM-DOOS* for *JEM-Distributed Objects Operating System*. Its inspiration owes a lot to POSIX[5], especially to POSIX.1. We adapt the way this norm deals with file systems to deal with object systems, i.e. hierarchies of objects similar to POSIX hierarchies of files. In the current release, *alpha 0.1*, objects we have implemented provide access to system resources, such as processors, screens, etc. Furthermore, *JEM-DOOS* supports remote access to objects, which makes it distributed. Hence, *JEM-DOOS* provides a way to deal with heterogeneous objects in heterogeneous networks of computers.

1 Introduction : The JEM Project

The *JEM* project[2] carried out at LaBRI has two main aims. The first is to facilitate the use and programming of distributed heterogeneous systems by means of the design and implementation of a distributed platform. Java[1] technology makes it possible for us to handle heterogeneity, and CORBA[4] and RMI[3] technologies makes it possible to deal with distribution, using remote method invocation and object transfer through the network. The current implementation is based on RMI. Future releases will also provide a CORBA interface.

The second aim is to use the above platform as a laboratory to study, with other research teams, some difficult problems related to parallel and distributed technologies[2]: algorithmic debugging; threads and their use, their debugging,

* Java and all Java-based marks are trademarks or registered trademarks of Sun Microsystems, Inc. in the United States and other countries. The author is independent of Sun Microsystems, Inc.

** This work is partly supported by the Université Bordeaux I, the Region Aquitaine and the CNRS.

their modelization and the problem of distributed priorities; mobile agents, their modelization and the validation of applications using them.

This paper describes the basis of our platform, *JEM-DOOS, Experimentation environMent for Java – Distributed Objects Operating System*. Its conception is similar to that of POSIX. POSIX defines a standard operating system *interface* and environment based on the UNIX operating system documentation to support application portability at the source level[6, 5]. The aim of *DOOS*, our Distributed Objects Operating System, is to provide _object[1] handling in a way similar to file (and directory) handling provided by POSIX. Note that we do not claim conformance to POSIX; we just applied a POSIX-like methodology to the design of our system. An _object of our system can be any object complying with a given interface (that we will define later in this paper). For instance in the current implementation we have developed _objects that give access to system resources through this standard interface. These _objects are `Machine`, `ProcessorManager`, `Processor`, `Screen`, etc. As a result, listing the contents of `/net/alpha.labri.u-bordeaux.fr/processors/` under *DOOS* gives a list of the processors of `alpha.labri.u-bordeaux.fr`.

The rest of this paper is organized as follows. We first give an overview of related work in Sect. 2. Section 3 presents the basic concepts of POSIX.1 related to files. Section 4 shows the design of *DOOS* compared to POSIX and Sect. 5 gives information about the current implementation compared to UNIX implementation of POSIX. Section 6 explains how Java/RMI technology makes *DOOS* a distributed system. We eventually provide an example application using *DOOS* in Sect. 7 and conclude with future directions of the project in Sect. 8.

2 Related Work

Many research projects are being developed around Java and distributed/parallel technologies. In this section we present some of the most significant of these projects.

Communication libraries. Communication libraries have been integrated into Java. jPVM[7] uses the possibility to interface native code with Java so as to make the PVM library available from Java. Another library, JPVM[8], is a PVM-like implementation totally written in Java. This implementation has the advantage not to require PVM on the target machines, and consequently to provide for a greater level of portability of applications. Other projects, such as E[9] (*The E Extensions to Java*), offer higher level extensions to basic Java communication mechanisms.

Extensions for parallel and distributed programming. Some systems propose to add new constructs for parallel programming, like Java //[10] for instance, or to add *frameworks* for data-parallel programming like Do![11]. Other projects, like JavaParty[12], offer both an execution support and new constructs to make easier the design and implementation of distributed Java applications.

---

[1] We use the notation _object to prevent any confusion between _objects handled by *DOOS* and objects in the usual meaning of object oriented technology.

Platforms. MetaJava[13] is a more general system. It relies on a meta level that makes it possible to manage a set of objects. It is based on a transfer of control to this meta level by events generation at the objects level. Hence, the meta level can coordinate, for instance in a distributed environment, the activities of the basic objects. Legion[14] is also a high level system. It provides a platform that integrates all the support necessary to manage a set of software and hardware distributed resources.

Our platform mainly differs from Legion in that it proposes a common interface to all objects of the system. See [15] for further comparison to other systems.

3 POSIX

To handle a file from within an application POSIX defines four entities: a file, a file descriptor, an open file description and an open file. These definitions lead to an abstract model of a possible POSIX implementation as shown in Fig. 1.

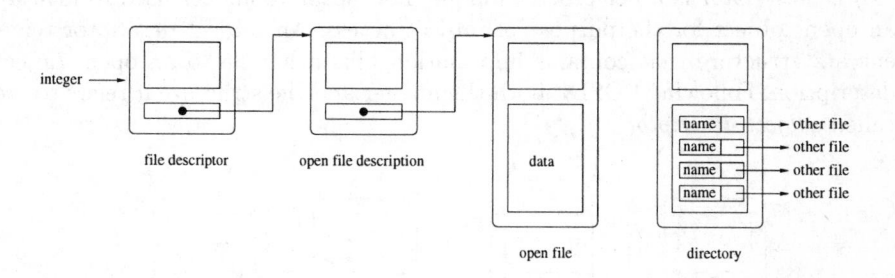

Fig. 1. A model of POSIX implementation and parenthood handling

A *file* is an object that can be written to, or read from, or both. An *open file* is a file that is currently associated with a file descriptor. An *open file description* is a record of how a process or group of processes are accessing a file. An open file description contains information on how a file is being accessed, plus a handle to an open file. A *file descriptor* is a per-process unique, non negative integer used to identify an open file for the purpose of file access. A file descriptor references a structure that contains information plus a handle to an open file description. Most of the time the name *file descriptor* is used for both this integer and the structure it refers to. The reason why this integer is required is for inheritance purpose between processes. Since processes do not share the same address space, using a handle, i.e. an address, would prevent them from inheriting/sharing file descriptors.

To ensure some sort of organization of files, POSIX provides the notion of directory (see Fig. 1). A directory is a file, the data of which are references to other files.

138

4 DOOS

We explained in Sect. 1 that the architecture of *DOOS* is closely based on the
POSIX file system architecture, but applied to objects: POSIX IOs deal with
files, i.e. *"dead" data* stored on devices, *DOOS* IOs deal with *"living" data*, i.e.
objects stored in memory. Hence, a *DOOS* _object[2] is an object. Because of this
similarity of conception, *DOOS* relies on a set of basic definitions that can directly be derived from those of POSIX. An *open _object* is an _object that is
currently associated with an _object descriptor[3]. _objects composing an _object
system can be anything. The sole constraint is that they show (by means of a
presentation layer as shown Fig. 4) the same open _object interface, that will
make it possible to handle them through a set of basic generic operations. For
instance, in the current implementation, we have developed _objects that provide
access to effective system resources such as files, screens, machines, processors,
etc. Of course, it is also possible to define _objects without relationship to system
resources. An *open _object description* is a record of how a process or group of
processes are accessing an _object. An open _object description contains information on how an _object is being accessed, plus a handle to an open _object. An
_object descriptor is a per-process unique, non negative integer used to identify
an open _object for the purpose of _object access. An _object descriptor references a structure that contains information plus a handle to an open _object
description. Following POSIX, both the integer and the structure it refers to are
called *_object descriptor*.

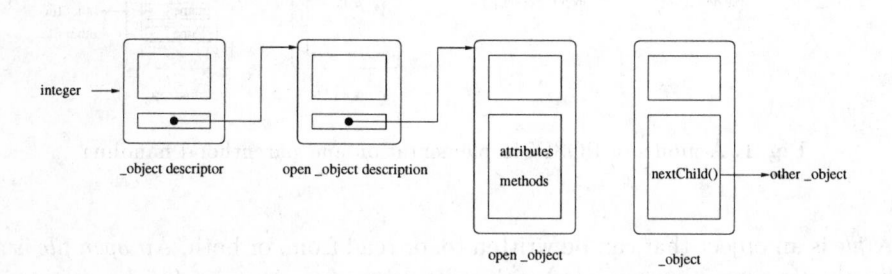

Fig. 2. A model of *DOOS* implementation and parenthood handling

DOOS provides a notion of parenthood between _objects. In the same manner
as a directory can contain references to other files, an _object has a method to
get handles to other _objects (see Fig. 2). This feature provides a hierarchy of
_objects in *DOOS*.

---

[2] Remember that we use the notation _object to prevent any confusion between
_objects handled by *DOOS* and objects in the usual meaning of object oriented tech-
nology. The distinction is not important for this precise definition, but will make
sense in the rest of this paper.

[3] When there is no possible confusion, we will use the term _object to talk about an
open _object .

The interface of a *DOOS* _object defines the following three features. *Naming:* `String getName()`: In the current release names are directly associated with _objects, not with parent _objects. Doing so, names can be built by _objects so as to describe the features they provide, e.g. the resource they interface (see Sect. 4). Names are mainly used for user interface purpose, for instance when listing the children of a given _object (see Sect. 7).

Input: `void copyIn(OpenObject object)` (POSIX `write`): Any _object can be copied into any _object . No global semantics is given to this operation. It is up to each _object of the _object system to decide how it deals with an _object copied into it. For instance, in the current implementation, an _object copied to a Screen _object is simply printed out and when a File _object is copied into another File _object , its contents replaces the contents of the original one.

Access to Children: The set of children of a given _object can vary during execution. Therefore, three functions are provided to access them: `startChildren` (POSIX `rewinddir`), `hasMoreChildren`, and `nextChild` (POSIX `readdir`). We cannot use a standard Java `Enumeration` since it would prevent the _object from building its children dynamically while they are enumerated.

This leads to the following Java interface for an open _object :

```
public interface OpenObject{
public String getName() throws labri.jem.exception;
 [...]
public void startChildren() throws labri.jem.exception;
public boolean hasMoreChildren() throws labri.jem.exception;
public OpenObject nextChild() throws labri.jem.exception;
public void copyIn(OpenObject _object) throws labri.jem.exception;
 [...]
}
```

Most *DOOS* _objects are lightweight since the above operations are usually straightforward to implement.

5 Low Level Implementation in Unix and DOOS

Under Unix, the fact that data stored in a physical device are interpreted as a file system, i.e. as files and directories, is the responsibility of both a layer of the operating system (which we call *presentation layer*) and a device driver (see Fig. 3). The presentation layer to be used depends on the device and file type. This layer is in charge of formatting physical data to provide the POSIX abstraction. Within this framework, a `readdir` operation could be implemented as follows (see Figs. 1 and 3):

1. [user] call `readdir(`*file descriptor* `fd)`
2. [system] access *open file description* to check access mode
3. [system] access *open file* to get physical data location
4. [system] get device driver and feed it with a read bytes request

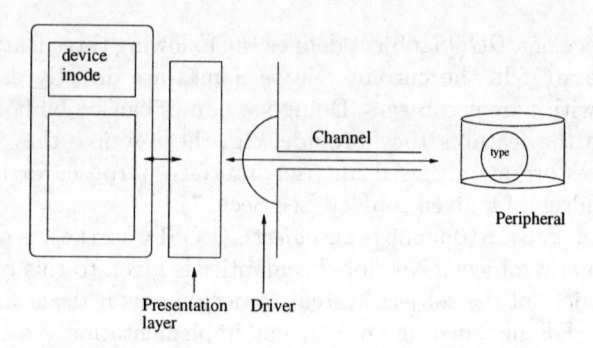

Fig. 3. A possible POSIX/UNIX implementation

5. [driver] apply operation
6. [system] format device driver answer depending on device and on file type
7. [system] return result to user
8. [user] use the result

In *DOOS* we do not need drivers as required in a UNIX implementation such as the one shown Fig. 3. The work which is done by a driver under UNIX mainly consists in transferring bytes between a physical device and the system. Under *DOOS* this is directly achieved by the Java Virtual Machine (Fig. 4). For instance, accessing the name of an object o is done by accessing o.name, what is effectively done by the JVM. Within this framework, a nextChild operation

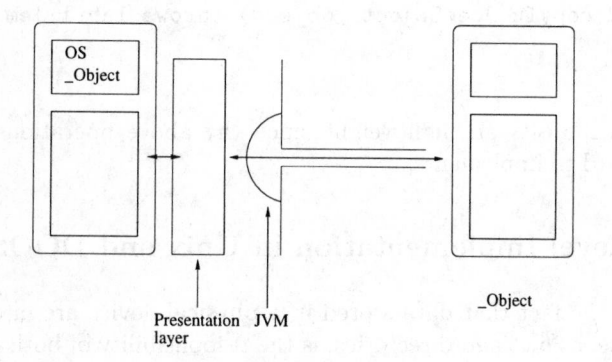

Fig. 4. *DOOS* implementation

(POSIX readdir) is implemented as follows (see Figs. 2 and 4):

1. [user] call nextChild(_object descriptor od)
2. [system] access *open _object description* to check access
3. [system] access *open _object* to get physical _object handle
4. [system] invoke nextChild() method on presentation layer associated to _object

5. [system] return result to user
6. [user   ] use the result

Each _object type is provided with a specific presentation layer (see Sect. 5). This layer is in charge of showing the OpenObject interface to *DOOS*.

6   The Distributed Architecture of *DOOS*

DOOS provides a way to aggregate a set of _object systems into a single distributed _object system. To do that it mainly relies on the possibility to have handles to remote objects. It is then enough to use remote handles instead of local handles in the structure presented in Fig. 2. These remote handles are implemented using the RMI framework provided by Java. RMI also provides distributed garbage collecting.

The basic feature required to make *DOOS* effectively distributed is the possibility to mount an _object system onto another _object system. This is illustrated by the following code. Although it will not be detailed here, it mainly consists in copying an _object to be mounted into a remote _object the mount point:

```
public static void main(String args[]){

OpenEntry myroot=
      new OpenEntry(null, args[0], new PMachine(null, args[0]));
OpenEntry mountPoint=DOOS.open("/net/alpha.labri.u-bordeaux.fr");
mountPoint.copyIn(myroot);
}
```

7   Example

The following example shows the implementation under *DOOS*, working at the open _object level (see Fig. 2), of a command equivalent to the Unix ls command:

```
public void list(String path){

  OpenObject oo = DOOS.open(path);
  oo.startChildren();
  while (oo.hasMoreChildren){
      OpenObject child = oo.nextChild();
      System.out.println(child.getName());
      DOOS.close(child);
  }
  DOOS.close(oo);
}
```

The major functional difference with the Unix ls command is that the *DOOS* command can be applied to any _object and it works both with local and remote _objects.

8 Conclusion and Future Work

We have presented version *alpha 0.1* of the *JEM-DOOS* Distributed Objects Operating System. We emphasized the fact that it is built on the same concepts as the UNIX file handling system, as defined by POSIX.1. We have introduced remote references to _objects based on the Java/RMI mechanism. We illustrated the use of the current implementation of the system.

JEM-DOOS is still evolving and more testing is required. Future work is concerned with the implementation of features for which we already provide appropriate structures, without using them. These are for instance access rights. In the longer term, we intend to provide a POSIX.2-like interface to final users of *JEM-DOOS*.

References

[1] K. Arnold and J. Gosling. *The Java programming language*. Addison-Wesley, 1996.

[2] S. Chaumette. Experimentation environMent for Java – Un environnement Java distribué pour le développement et l'expérimentation. Technical Report 1207-98, LaBRI, Mai 1998.

[3] D. Lea. *Concurrent Programming in Java*. Addison-Wesley, 1997.

[4] J. Siegel. *CORBA, Fundamental and Programming*. Wiley, 1996.

[5] *POSIX Part 1: System Application Program Interface (API) [C Language] ISO/IEC 9945-1 (ANSI/IEEE Std 1003.1, 1996 Edition)*.

[6] J.P. Braquelaire. *Méthodologie de la programmation en C – Bibliothèque standard - API POSIX*. Masson, 1998.

[7] jPVM project web page, 1998. http://www.isye.gatech.edu/chmsr/jPVM/.

[8] A.F. Ferrari. JPVM: Network parallel computing in Java. Technical Report CS-97-29, Department of Computer Science, University of Virginia, Charlottesville, USA, December 1997.

[9] E, The E Extension to Java, project web site, 1998.
http://www.communities.com/products/tools/e/e_white_paper.html.

[10] J. Vayssière. Programmation parallèle et distribuée en java. conception et implémentation de java //. Rapport de DEA et de fin détudes, projet SLOOP, I3S, Université de Nice, INRIA Sophia-Antipolis, Septembre 1997.

[11] P. Launay and J.L. Pazat. Generation of distributed parallel java programs. Technical Report 1171, IRISA, Fevrier 1998.

[12] M. Philippsen and M. Zenger. JavaParty – Transparent Remote Objects in Java. *Concurrency: Practice and Experience*, 9(11):1225–1242, November 1997.

[13] M. Golm and J. Kleinöder. MetaJava – a platform for adaptable operating-system mechanisms. Technical Report TR-14-97-10, Computer Science Department, Friedrich Alexander University, Erlangen-Nürnberg, Germany, April 1997.

[14] A.S. Grimshaw, A. Wulf, and the Legion team. Legion. The next logical step toward the world-wide virtual computer. *Communications of the ACM*, 40(1), 1997.

[15] S. Chaumette. JEM experimentation environMent for Java compared to other objet systems. Technical report, LaBRI, 1998. To be published.

Static Networks: A Powerful and Elegant Extension to Concurrent Object-Oriented Languages

J. Yelon and L. V. Kalé

University of Illinois, Urbana Illinois, USA

Abstract. A network of objects is a set of objects interconnected by pointers or the equivalent. In traditional languages, objects are allocated individually, and networks of objects are assembled incrementally. We present a set of language constructs that can create *static networks*: networks of objects which are created atomically, and which are immutable. Then, we present some of the most interesting abilities of static networks.

1 Introduction

In most concurrent object-oriented languages, objects are allocated individually, and networks of objects are assembled incrementally. We have developed a new way to allocate networks of objects. When using our constructs, entire grids, trees, or graphs pop into existence instantaneously. The structures created this way are immutable. Such networks are termed *static networks*. We have discovered that static networks lead to more elegant code and better software engineering. For example, we have discovered that static networks make it possible to define a new kind of lexical scoping mechanism. We have also discovered that static networks can dramatically reduce the amount of code needed to interface objects to each other.

2 The Constructs

Our static network construct could be added to any existing concurrent object-oriented language. We have been using a variant of Java called "Distributed Java" for our experiments. It is a very conventional concurrent object-oriented language, we chose it for its simplicity and familiarity. It contains the usual constructs of Java. To this, we added several conventional concurrency constructs.

Our first addition to Java is the operator *new classname(constructorargs) on processor*, that allocates an object on a specified processor. It returns a *proxy* of the object: a tiny object that masquerades as the real object. When one invokes a method on the proxy, the arguments are transparently shipped to the real object, and the return-value transparently shipped back. If the object is remote, the method receives copies of the parameters which are indistinguishable from the originals. If the parameters are of numeric types, this is easy. If the parameters are of object types, the callee receives a proxy. It is often desirable, when passing data across processor boundaries, to copy the data instead of creating a proxy. Therefore, we add a parameter-list keyword *copy* that allows the system to copy the parameter instead of provide a proxy.

Java already has threads. We added a shorthand form of the thread creation mechanism: *object<-method(arguments)*. The new thread invokes a method on an object and then terminates. Unlike some concurrent object-oriented languages, we allow multiple threads to execute within the same object at the same time. However, any contiguous sequence of statements is atomic unless it contains a *blocking* statement. There are two blocking statements: method invocation, and the *wait* statement. The *wait* statement causes the current thread to suspend. Any modification to the instance variables of the current object awakens the thread. In other words, it means *wait for an instance variable to be modified*. There is a second form of the *wait* statement: *wait condition*. It checks the condition, and if false, it suspends. When reawakened by the modification of an instance variable, it rechecks the condition, and if still false, it resuspends.

The operation *newgroup classname(constructorarguments) [size]* creates a one-dimensional array of objects that spans multiple processors. It returns the handle for the group. One can invoke a method on any element of the group by saying *grouphandle[index].method(arguments)*. One can invoke a method on all elements of the group by saying *grouphandle[ALL]<-method(arguments)*. One can invoke a method on an arbitrarily-selected member of the group by simply saying *grouphandle.method(arguments)*. The latter behavior emulates the behavior of Concurrent Aggregates [1], HAL groups [2], and Charm++ branch offices [3]. It enables a group to act as a bottleneck-proof object. Any member of the group can determine its position in the group by evaluating the pseudovariable *thisindex*. Any member of the group can obtain the group handle by evaluating the pseudovariable *thisgroup*.

2.1 Static Object Hierarchies

Now that we have a simple foundation language, we add the static network support. The core of our static network support is the *agent* declaration. The agent declaration is most easily explained by comparing it to some traditional code with a similar effect:

```
class Ct {
  D T[10];
  Ct() {
    for (int i=0; i<10; i++)
      T[i] = new D(1000);
} };
```
```
class C {
  agent T(int i) is D(1000);
};
```

Class *Ct* is traditional code. Each time an object of class *Ct* is allocated, 10 objects of class *D* automatically pop into existence (because of the constructor). Those objects are named *T[0]*, *T[1]*, etc. When you allocate an object of class *Ct*, you're effectively allocating a small tree of objects, with the object of class

Ct at the root, and the 10 objects of class *D* underneath. Class *C* uses the agent declaration to achieve an almost-identical effect. Each time an object of class *C* is allocated, a conceptually infinite set of objects of class D pop into existence. Those objects are named *T(0)*, *T(1)*, etc. They are allocated lazily the first time they are accessed. So once again, allocating an object of class *C* effectively allocates a small tree of objects.

More generally, the syntax of the agent declaration is is *agentname(indices) is classname(constructorargs)*. It must always occur inside a class declaration. Assuming it occurs inside a class *C*, it declares that each object of class *C* has several "agents" (other objects working for it). The object of class *C* can access its agents by evaluating the expression *agentname(indices)*, and the system ensures that they are there when accessed. The agents are of class *classname* and are initialized with the specified *constructorargs*. The *constructorargs* can refer to the agent indices and to the constructor arguments of class *C*. There can be as many indices as are desired, they can be any type which could reasonably function as a hash table key.

The real difference between using a constructor, as in *Ct*, and an agent declaration, as in *C*, is that *C*'s hierarchy is immutable. It also presents the illusion of having been created atomically.

2.2   Communication Across the Hierarchy

The agent construct can only create trees of objects. However, if we allow sibling-to-sibling communication, then the hierarchies can also emulate other structures. Consider, for example, a class *jacobi* containing an agent declaration *agent jnode(int i, int j) is jacobinode()*. The agents form a shallow hierarchy with a single *jacobi* object in charge of innumerable *jacobinode* objects. If *jnode(i,j)* invokes methods on its siblings *jnode(i+1,j)*, *jnode(i-1,j)*, *jnode(i,j+1)*, and *jnode(i,j-1)*, then the lines of communication form a grid. So this agent hierarchy is in one sense a tree, but in another equally-meaningful sense it is a grid. Because we want to be able to represent arbitrary structures like grids, trees, graphs, and other communication patterns, it is important that the objects in an agent hierarchy be able to send messages to their siblings.

If you implement hierarchies of agents using parent pointers and child pointers, then sibling-to-sibling communication cannot be done. The siblings don't have pointers to each other. Because of this, we use a completely different implementation procedure for agents. We assign each object an ID string according to two simple rules. There are two disjoint kinds of objects: those allocated with *new*, and those which are agents. The ID of an object allocated with *new* is of the form *classname#seqno@processor*, where the *seqno* is a unique ID to tell objects apart, and *processor* is the location of the object. The ID of an object which is an agent is of the form *ownerid.agentname(indices)*, where *ownerid* is the ID string of the agent's owner, and *agentname(indices)* is the agent's name and indices as they appear in the code. An object in the agent hierarchy can easily use string manipulation to compute the ID of a sibling, parent, or child. IDs are mapped statically to processors, and every agent is stored in a hash table (by ID) on its home processor.

When remotely invoking a method on an agent, one must perform four steps that are not necessary for other objects. On the sender side, one must compute the ID of the agent, one must compute its hash value, and one must determine its home processor by mapping its ID to a processor number. At the receiver side, one must look the object up in the hash table. We picked a typical statement that involves all these steps from one of our demonstration programs and hand-optimized it. We executed the code on a Cyrix PR266 processor, and discovered that the sender-side steps took 0.5 microseconds, and the receiver-side steps took 0.2 microseconds. It is our judgment that these values are within the acceptable range. Note that our prototype implementation does not yet generate code of this quality.

The user of Distributed Java never sees an agent ID, they are hidden inside proxy objects. One utilizes a variety of notations to obtain proxy objects. The simplest is when an object X evaluates the expression *agentname(indices)*. The system implicitly concatenates ".agentname(indices)" to the ID of X, hides the resulting ID inside a proxy object, and returns the proxy. Another notation is when an object X evaluates the keyword *owner*. The system implicitly truncates the ID of X, yielding the ID of X's parent in the hierarchy. Again, it hides the ID in a proxy and returns the proxy.

Getting the proxy of a sibling is implementationally easy, but it would break scoping rules. Consider class C in Section 2.1. Class C contains the declaration of the T agents. So it is reasonable to refer to those agents by name (e.g. $T(0)$, $T(1)$, etc.) inside the methods of C. But it is not reasonable to refer to the name T in the methods of D, since the declaration of T isn't anywhere near D. There are two solutions. The first is to move the declaration of D inside class C, as shown below. This puts the code of D into a scope where the T agents are visible. That way, the methods of class D can straightforwardly refer to their siblings by name.

```
class C {
    agent T(int i) { /* insert body of class D here */ }
};
```

The second solution is used when you can't move D into C, for example, when D is a library class. In that case, the solution is to program D so that it sends its output to its owner. The owner can then forward the data to wherever it needs to go. The data goes from D to C to D. To avoid a bottleneck at the C object, we provide *relay methods*. One places the keyword *relay* in front of a method declaration. The compiler prohibits relay methods from accessing instance variables. But since the relay method doesn't actually access the object, it can be performed on any processor. Invocation of a relay method does not cause a remote method invocation, the method is performed locally. To use this in our sample problem, the object of class D invokes a relay method in class C, which does nothing except invoke a method on another D. At the implementation level, only one remote method invocation happens, going from D to D.

2.3 The From Clause

We now add one more feature. We allow an object to ask: "who invoked this method?" At a glance, this may not appear to be related to static networks. However, if one were to ask this question in a traditional language, the only answer one could expect would be something like "you were invoked by object #5907." That answer doesn't do you any good. Static networks make it possible to get an answer of the form "this method was invoked by agent $T(5)$", which is a meaningful and useful answer. So effectively, static networks are what make the *from* clause possible. Its notation is shown in the sample code below:

```
class C {
    agent T(int i) { ... owner<-data(f(g(i))) ... }
    public void data(int n) from T(int j) {
        printf("T(%d) sends the result %d\n",j,n);
} };
```

In this sample, each agent *T(k)* computes a value and sends it back to its owner, the object of class C, by invoking the *data* method on it. If an agent *T(k)* invokes *data*, then the *from* clause will cause *data* to execute with x bound to k. If any other object tries to invoke *data*, this definition of the method will not be visible. However, method overloading is permitted as usual, so there may be other definitions of *data*, one of which may be visible.

3 Shared Variable Scoping

One surprising ability of static networks is their ability to create lexical scoping rules for shared data. Some concurrent object-oriented languages do provide global variables, which are shared. However, if an algorithm stores data in a global variable, you typically can't run multiple copies of that algorithm at the same time. In other words, the use of global variables tends to make your code nonreentrant. This limits the utility of global variables in a parallel program, where the whole idea is to run many copies of many algorithms concurrently. Because of these reentrancy problems, many concurrent object-oriented languages do not provide shared variables at all. But shared variables are quite useful: it's their global scope which is the problem.

Agents can provide shared variables which *aren't* global. Consider the code below. The left side shows the prototype of a *writeonce* class which holds a single integer. It provides methods *set* and *get* which can set and retrieve the integer. The *get* method is supposed to block until the value has been *set*. The right side shows class *K*. The object of class *K* will own a *writeonce* object named *V* and a number of objects named *T(0)*, *T(1)*, etc.

class writeonce {	class K {
void set(int n);	agent V is writeonce();
int get();	agent T(int i) { /* these objects may refer to V */ };
};	};

We can define the word *variable* to mean a storage location named by an identifier, where the identifier has a scope. By that definition, the agent *V* is a variable. It is accessible to several objects *T(0)*, *T(1)*, etc, so it is a shared variable. Its scope is not global. If I create a two instances of class *K*, there will be two copies of this hierarchy, and each will include its own copy of *V*. The two hierarchies will function independently of each other. In short, static networks are giving us true static scoping. This in turn makes it possible to use shared data without sacrificing reentrancy.

4   Improved Compositionality

Concurrent object-oriented languages make it extremely difficult to interface concurrent objects to each other. To demonstrate the problem, we will attempt a programming task that should be easy, but turns out not to be.

The task is to start with the two library classes shown below, and compose them in such a way that that they perform the computation $A * B * C$, where A, B, and C are square matrices. *matmul* below is a simple matrix multiplier, one feeds in the rows of A and columns of B, and it sends out the elements of the result matrix. *conv2rows* is a converter that rearranges the elements of a matrix: one feeds in columns or individual elements, and it produces rows. Both classes use continuation-passing style (CPS). In other words, they accept the handle of a group *sendto*, and they produce their output by sending it directly to the specified group. We chose continuation-passing style because it is the only efficient interface for these classes. Using call-return communication would double the bandwidth requirements or introduce unnecessary bottlenecks. CPS is common in concurrent object-oriented programs.

class matmul {	class conv2rows {
matmul(int rows, int cols, object sendto);	conv2rows(int rows, int cols, object sendto);
void row_a(int r, vector v);	void col(int c, vector v);
void col_b(int c, vector v);	void elt(int r, int c, double v);
};	};

The design of the code is as follows. We will create a group *mulabc* to be the coordinator. For consistency, it will expect all the inputs A, B, and C to be fed in one column-vector at a time. The *mulabc* group will create two *matmul* groups, one to multiply $(A * B)$, and one to multiply $(A * B) * C$. It will create a group *conv2rows* to convert the columns of A into the rows of A, and another *conv2rows* to convert the elements of $(A * B)$ into the rows of $(A * B)$.

Though the first *matmul* group has a *sendto* parameter, I cannot simply configure it to send its output to the first *conv2rows*. The problem is that the method names don't match: *matmul* wants to invoke the *result* method, but *conv2rows* wants its user to invoke *input_elt*. The solution is to configure the *matmul* such that it sends its output back to the *mulabc*. The *mulabc* then forwards the data back to the *conv2rows*, using the right method name. In fact, we must use the same strategy everywhere: all groups will send their intermediate results back to the *mulabc* for dispatching.

CPS now creates a problem: the *mulabc* cannot tell the output of the first matrix multiplier from the output of the second multiplier. To distinguish them, we have to send the two outputs to two different places. So we must create a second group *mulabc1*, to give the second multiplier a place to which to send its output. This completes the design, two versions of the code are shown below:

```
class mulabc {
  matmul mm1, mm2; conv2rows cr;

  mulabc(int size, object sendto) {
    if (thisindex == 0) {
      intermed = newgroup mulabc1();
      mm1 = newgroup matmul(size, size, intermed);
      mm2 = newgroup matmul(size, size, thisgroup);
      cr  = newgroup conv2rows(size, size, intermed);
      er  = newgroup conv2rows(size, size, thisgroup);
      // broadcast object handles to all members of group.
      intermed[ALL]<-distribute_handles(mm1, er);
      thisgroup[ALL]<-distribute_handles(mm1, mm2, cr);
    }
  }

  public void distribute_handles(matmul mm1_a,
        matmul mm2_a,conv2rows cr_a) {
    mm1 = mm1_a;
    mm2 = mm2_a;
    cr  = cr_a;
  }

  void input_A(int col, vector v) {
    wait (cr != NIL);
    cr<-col(vol, v);
  }
  void input_B(int col, vector v) {
    wait (mm1 != NIL);
    mm1<-col_b(col, v);
  }
  void input_C(int col, vector v) {
    wait (mm2 != NIL);
    mm2<-col_b(row, v);
  }
  void row(int row, vector v) {
    wait (mm2 != NIL);
    mm2<-row_a(row, v);
  }
  void result(int row, int col, double d) {
    wait (sendto != NIL);
    sendto<-result(row, col, d);
}};
```

```
class mulabc1 {
  matmul mm1;
  conv2rows er;
  void distribute_handles
            (matmul mm1_a, conv2rows er_a) {
    mm1 = mm1_a;
    er = er_a;
  }
  void row(int row, vector v) {
    wait (mm1 != NIL);
    mm1<-row_a(row, v);
  }
  void result(int row, int col, double d) {
    wait (er != NIL);
    er<-elt(row, col, d);
}};
```

```
class mulabc {
  mulabc(int size) { };
  agent mm1 is matmul(size, size);
  agent mm2 is matmul(size, size);
  agent cr is conv2rows(size, size);
  agent er is conv2rows(size, size);
  relay void input_A(int col, vector v) {
    cr<-col(col, v);
  }
  relay void row(int row, vector v) from cr {
    mm1<-row_a(row, v);
  }
  relay void input_B(int col, vector v) {
    mm1<-col_b(col, v);
  }
  relay void result(int row, int col, double d) from mm1 {
    er<-elt(row, col, d);
  }
  relay void row(int row, vector v) from er {
    mm2<-row_a(row, v);
  }
  relay void input_C(int col, vector v) {
    mm2<-col_b(row, v);
  }
  relay void result(int row, int col, double d) from mm2 {
    owner<-result(row, col, d);
}};
```

The version using agents (inset) is about half the size of the traditional version. The complexity in the traditional version comes from four sources. First, *mulabc* has to be a group to avoid bottlenecks. Agents eliminates the bottleneck by means of *relay* methods. Second, everything must be allocated, connected together, and pointers passed about: this takes a significant amount of code. Third, the traditional version contains much explicit synchronization to deal with the fact that the network is not created atomically. Agents creates the entire structure instantaneously, eliminating the need for explicit synchronization. Fourth, the *from* clause eliminates the problems created by *CPS*, which eliminates the need for *mulabc1*. The combined effect is dramatic.

5 Summary and Conclusions

We have identified a new way to allocate networks of objects: atomically. We developed a language-level construct, the *agent* declaration, that allows us to create hierarchies of objects atomically. Such hierarchies can emulate other structures like grids and graphs. We described two major advantages. One, they enable us to support static scoping rules, enabling the reentrant use of shared data. Two, they significantly simplify the interfacing of modules, improving the compositionality of the language. These are only two of the uses: a paper of this size is not sufficient to describe all the possibilities. We have discovered that agent ID strings make a powerful tool when used for program trace analysis. We have discovered that the agent construct subsumes the *group* construct, allowing us to eliminate it. This means that the language is not becoming more complex with the addition of agents. We have discovered that the zero-cost creation of agent structures makes it possible to set up an large structure and only use it once. This leads to a programming style that is more like functional programming. We do not expect to stop discovering new uses for static networks.

References

[1] A. Chien and W. J. Dally. Concurrent aggregates. In *Second ACM Symposium on Principles and Practice of Parallel Programming*, Mar 1990.
[2] G. Agha, W. Kim, and R. Panwar. Actor languages for specification of parallel computations. In *DIMACS Series in Discrete Mathematics and Theoretical Computer Science, Specification of Parallel Algorithms*, 1994.
[3] L. V. Kalé and S. Krishnan. Charm++: A Portable Concurrent Object Oriented System Based on C++. In *OOPSLA*, 1993.

A FIFO Queue Class Library as a State Variable of Time Warp Logical Processes

Soichiro Hidaka, Terumasa Aoki, Hitoshi Aida, and Tadao Saito

University of Tokyo, Tokyo, Japan

Abstract. We describe SsQueue(SnapshotQueue), an implementation for an efficient and user-friendly class library for FIFO queue that can be used for state vectors in simulated queuing network constructs under Time Warp parallel discrete event simulation (PDES) protocol. There exists a general purpose Time Warp simulation kernel WARPED, where users have only to define a state vector and do not have to care about rollback and state recovery. However, since the state vector should be defined as an inlined data structure, it is not suitable for dynamic data structures such as FIFO queue. This class can also serve as an element of such a state vector, then both libraries and users can handle each instance as snapshot of the queue. Taking advantages of FCFS nature of the above data structure, operation histories rather than all contained items can be safely stored and restored using this class library with virtually minimum overhead. When the kernel deletes instances in the simulated past, corresponding methods perform garbage collections transparently.

1 Introduction

Discrete event simulation (DES) technology has been playing an important role in design and evaluation of computer and communication systems. However, rapid growth of these systems in both scale and speed have made DES relatively slower. Large scale, high-speed Asynchronous Transfer Mode (ATM) networks and its components such as multi-stage switches are among these systems. We can benefit from parallel processing provided that simulation models of such systems have ample parallelism. In this case, simulation models are divided into sub-models called *Logical Processes* (LP). Each LP maintains its own simulation clock and internal state, propagating any causal effects to other sub-models using external event messages. Many parallel discrete event simulation (PDES) protocols to synchronize LPs have been proposed, which are roughly divided into two categories. In *conservative* protocols[1], each LP strictly adheres to causality constraints. In *optimistic* protocols[2], each LP proceeds speculatively. When a causality error occurs, it rollbacks, restores its state, and resumes simulation.

In optimistic protocols, state saving is necessary to rollback and coast-forward along simulated time. However, state saving and rollback are too complex and error prone for application programmers to implement by themselves.

Although there are several simulation libraries and languages that support these mechanisms automatically[3, 4, 5, 6], they all adopt fixed-size inlined data structures (e.g. integers, floating-point numbers) as the elements of their state vectors. They are not suitable for dynamic data structures such as queues in queuing network simulations.

There might be, of course, some ways to implement queuing data structure using inlined data structures. For instance, we can use LPs to simulate each container of a queue (Fig. 1(a)). Such a method may be most suitable for hardware simulation practitioners who are interested in detailed behaviors of queues themselves. However, for designers who are more interested in higher-level simulations where queues are viewed merely as FIFO buffers, the above method is both too complicated and inefficient. Furthermore, since there is no "Global State" in PDES, information regarding to which LP has the next candidate of "enqueue" and "dequeue" must be exchanged using additional external event messages.

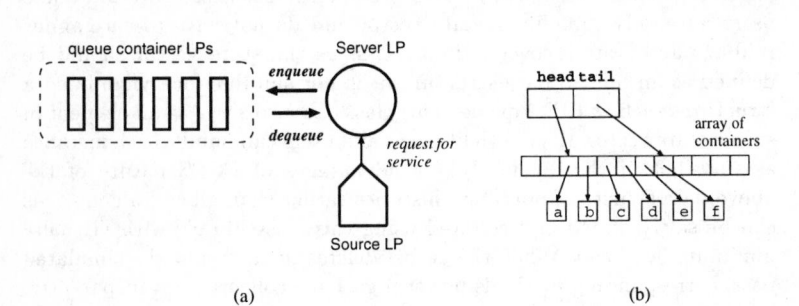

(a)   (b)

Fig. 1. Alternative implementation strategies for FIFO queue; (a) associating each container with an independent LP and (b) using an array of containers as a ring buffer

Another way to implement a queuing data structure as an element of a state vector using inlined data structure is to use an array of containers (pointers) whose length is equal to the maximum number of items to be inserted (Fig. 1(b)). Although this method is easy to implement, the memory consumption and consequently the cost of each state saving is always proportional to the array length.

In addition, both of the above techniques must fix the maximum number of items that can be inserted a priori. This constraint is burdensome when one wants to determine by simulation how long should be the lengths of queues in switches to achieve a required upper bound in loss rate. In such a simulation, the maximum instantaneous queue length may be of interest.

This problem can be solved using incremental state saving (ISS) facilities in SPEEDES[7]. In using IIS, not all elements themselves have to be stored in order to accomplish state saving of such a "redundant" data structure.

The SPEEDES kernel also maintains rollback and coast-forward automatically, but it performs user-provided DO/UNDO operations in the state queue rather than copying-in and copying-out of the corresponding snapshot in the state queue as in WARPED[3].

Any operations that can be undone are successfully maintained using ISS. However, when a FIFO queue is implemented using IIS, the advantage of its FCFS nature is not fully exploited. Consequently, the cost of rollback is proportional to the rollback distance since successive operations must be undone one by one.

In our approach, the cost of rollback is independent of the rollback distance (except for the cost of invalidating cancelled state history), because the kernel simply lets the application programmer access the appropriate snapshot instead of the current state.

The combination of our approach and WARPED exploits another benefit. The kernel may want to take snapshots of the state infrequently as an optimization. Our approach is independent of the frequency of state saving, as multiple enqueue/dequeue operations can be represented without taking snapshots each time. In IIS, on the other hand, each enqueue/dequeue operations require DO/UNDO data structures to be saved. That is to say, the frequency of state saving is equal to the frequency of enqueue/dequeue operations.

The rest of this paper is organized as follows. Section 2 gives an overview of our class library. Section 3 describes design and implementation as well as API of our class library in detail. Section 4 provides an example of Time Warp queuing network simulation using our class library. Final comments and concluding remarks are given in Sect. 5.

2 Overview of the Class Library

In a FIFO queue, unlike a priority queue, rearrangement of elements does not occur, so it is easy to implement "difference" based state savings. Common elements among snapshots (elements enqueued before the previous snapshot and not dequeued as of current snapshot) are shared via pointers. These mechanisms are encapsulated into the C++ class library implementation and are transparent both to application programmers and to the simulation kernel. Henceforth we call this package *SsQueue* (*Snap shot Queue*), since each of their instances represents a snapshot at the moment it are created. SsQueue achieves virtually minimum state saving and recovery overhead (each instance consists of five pointers and an integer) which is independent of the number of items inserted.

3 Implementation

This section explains the API from a programmer's point of view, the interface from the simulation kernel's point of view, and finally the internal representation.

3.1 API on User's Side

The interface on the user's side is quite simple.

- `void enqueue(void *item)` inserts an `item`.

- `void *dequeue()` removes and returns the first item.
- `void *top()` is a non-destructive version of `dequeue()`.
- `int getLength()` obtains current length.

3.2 Interface on Simulation Kernel's Side

The Time Warp simulation kernel performs periodic state saving by copying current state variables to the state queue. When rollback occurs, the snapshot just before the rollback point in the state queue is copied back to the current state variables. SsQueue traps both operations by providing a custom copy constructor and assignment operator. In addition, when the kernel "deletes" (memory resources occupied by committed states in the state queue should be returned to the system) an SsQueue instance, its destructor performs garbage collection of its own. Modification of the simulation kernel is not necessary in using SsQueue.

3.3 Internal Representation

Implementation of the temporal relationship and sharing elements among neighboring snapshots in SsQueue is shown in Fig. 2. Application of enqueuing and dequeuing operations to the current state ($S(t + 1)$ in the figure) is straightforward. In enqueuing an item, the corresponding container is added at the tail of the container's list, and `tail` pointer is modified to point to that container. When an element is dequeued, the item pointed to by `head` is returned to the user, `head` moves on to the next container (in the figure, container that points to "g"). Note that dequeued items and their containers are not actually "deleted" immediately. They are deleted when the kernel deletes the corresponding snapshot as described below.

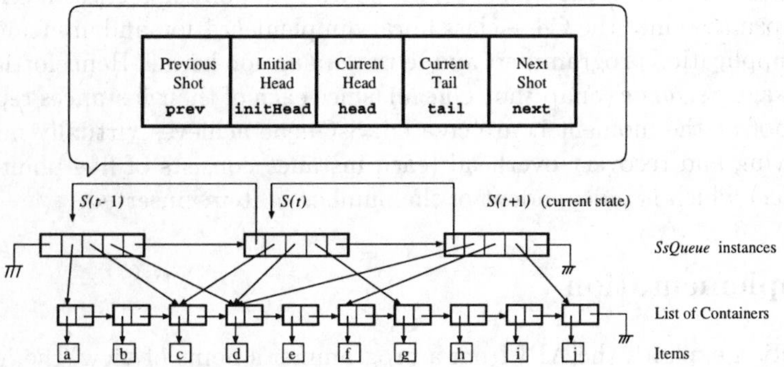

Fig. 2. Temporal relationship and sharing elements among neighboring snapshots in SsQueue instances

A temporal relationship is established when the kernel performs state savings and restorations. Overriding copy constructor and assignment operator trap assignment operations in both cases. When a state is saved, the assignment operator is called with source being the current state and target being a snapshot in the state queue. In this case, SsQueue links this new snapshot just before the current state (i.e., next of the new snapshot points to the current copy of SsQueue and prev of current copy points to the snapshot). On rollback, the kernel copies the latest committed snapshot back to the current state variables. SsQueue knows this fact since this time, the assignment operator is called with a snapshot as its copy source. This causes SsQueue to invalidate cancelled snapshots and corresponding items (they may be recycled before actually being deleted), and re-link the restored state as direct predecessor of the new current state.

Garbage collection is invoked from the overridden destructor when kernel deletes committed states. Take $S(t-1)$ in Fig. 2 as an example. Containers (and items) from inithead to just before head must be deleted. Items "a" and "b" are deleted in this case.

These constructs are transparent to both the simulation kernel and application programmers, letting the former remain unchanged, and the latter avoid handling of rollback and state recovery. The cost of state saving and restoration is reasonably small, as each snapshot requires only five pointers (prev, next, inithead, head, tail) and an integer (length). When a state is recovered, the selected state is always ready for subsequent enqueuing and dequeuing, yielding (other than invalidating descendants) constant and thus minimum overhead.

4   Example Using SsQueue

This section illustrates a Banyan switch simulation as an example of large-scale parallel queuing network simulations. A Banyan switch is a multi-stage self-routing switching network that is used in high-speed broadband network switches. Figure 3 shows how an 8 port, 3 stage Banyan switch is configured. Each 2×2 switching element (inside round rectangle) reads a header in an incoming ATM cell, and selects an output port. The cell is buffered if it attempts to go to a port in use. All of these operations together form an 8×8 switch, performing point-to-point routing of ATM cells from input ports to output ports.

The simulation program is build on top of WARPED[3], a public domain Time Warp simulation kernel written in C++. In WARPED, users define their own LPs by inheriting the basic LP class[1], defining a custom event processing method which is called by the kernel. The basic LP class is provided as a template class that takes a state vector class as its argument.

We defined four LPs to simulate a Banyan switch. They are CellSrcObj to generate input traffic, CellRouterObj, a 2×2 router, CellOutBufObj for output buffer and port, and CellSinkObj that collects cell statistics on each output port. Figure 4 shows how the SsQueue class is integrated into the CellOutBufObj user program.

---

[1] In WARPED, the term LP refers to a per-processor entity which aggregates multiple simulation objects. We call these multiple objects LP here by convention.

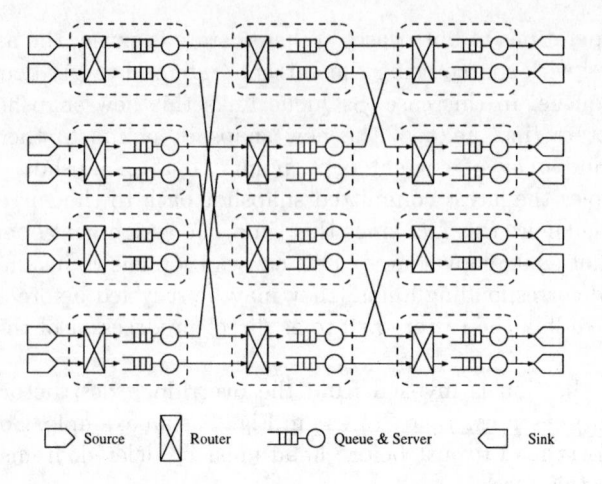

Source ▷ Router ⊠ Queue & Server Ⅲ-○ Sink ◁

Fig. 3. LPs constructing 3 stage Banyan switch

```
1:/************** State variables definition ********************/
2:class CellOutBufState : public BasicState {
3:public:  SsQueue ssq;    // SsQueue instance as a state variable
4:         int isBusy;     // 1 if server is busy
5:};
6:
7:/********************** LP definition ***********************/
8:class CellOutBufObj : public SimulationObj<CellOutBufState> {
9:public:  void executeProcess();  // user-defined event handler
10:  ...
11:};
12:
13:/**************** User defined event handler ******************/
14:void CellOutBufObj::executeProcess() {
15:  CellEvent *event = (CellEvent *)getEvent(); // next evt to exec
16:  switch(event->type) {
17:  case ARRIVECELL:          // Arrival of ATM cell
18:    SsQueue *ssq = &(state.current->ssq);
19:    if (!(state.current->isBusy) && ssq->isEmpty()) {
20:      // cell is forwarded immediately without queuing
21:    } else {   // buffer this cell
22:      CellInfo *cell = new CellInfo;
23:      *cell = ((ArriveCellEvt *)ace)->cell;
24:      ssq->enqueue(cell);  // cell is buffered into SsQueue
25:    }
26:    break; ...
27:  }
28:}
```

Fig. 4. Coding example using SsQueue

State vector declaration is given in lines 2 to 5, including SsQueue as part of state variables (line 3). LP is defined using this state variable, as in lines 8 to 11. Lines from 17 to 26 give an idea of how a cell arriving at `CellOutBufObj` is handled. If the output port is busy, the cell is put into a SsQueue instance, as in lines 19 and 24.

Performance evaluation of the whole simulation program is given as elapsed real time to simulate 50000 μseconds(Fig. 5). An experiment was carried out on 1 to 32 processors of SR2201[8], a distributed memory multicomputer. Relatively good speedup is achieved, especially when the number of switching elements (SEs) increases.

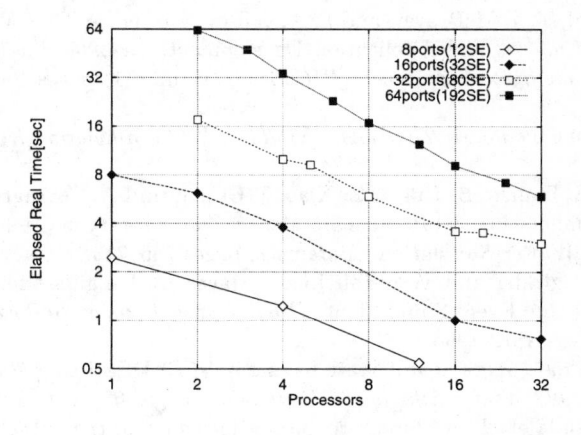

Fig. 5. Elapsed real time in seconds to simulate 50000 μseconds

5   Conclusion

In this paper, SsQueue, an efficient and user-friendly class library implementation of a FIFO queue for Time Warp simulation was described. Each instance of this class occupies a reasonably small memory space, so that state saving cost is independent of the number of items in the queue. Sharing containers among adjacent snapshots minimizes the cost of garbage collection.

SsQueue stands between the provider of a general purpose Time Warp simulation platform and application programmers. The simulation kernel needs only support inlined data as state variables of Logical Processes, although it may be responsible for all aspects of optimistic synchronization. Using SsQueue, simulation programmers who do not want to be concerned about rolling-back and state saving can still have the queue data structure as state variables of Logical Processes in such a simulation kernel.

Application to Banyan switch simulation was also presented, which showed both the simplicity of integration and the ability to be a part of real-world application programs.

Acknowledgements We would like to thank Dr. Yutaka Ishikawa in the RWCP for his valuable suggestions. We also thank the WARPED project contributors. If their simulation kernel were not in public domain, this paper would not have been possible.

References

[1] J. Misra. Distributed discrete-event simulation. *ACM Computing Surveys*, 18(1):39–65, 1986.

[2] D. A. Jefferson. Virtual Time. *ACM Transactions on Programming Languages and Systems*, 7(3):404–425, 1985.

[3] D. E. Martin, T. J. McBrayer, and P. A. Wilsey. warped: A Time Warp Simulation Kernel for Analysis and Application Development. In *29th Hawaii International Conference on System Sciences (HICSS-29)*, volume 1, pages 383–386, January 1996.

[4] University of Cincinnati. *The WARPED Time Warp Simulation Kernel*, July 1996. Ver 0.7.

[5] F. Gomes, S. Franks, B. Unger, Z. Xiao, J. Cleary, and A. Covington. SIMKIT: A High Performance Logical Process Simulation Class Library in C++. In *Proceedings of the 1995 Winter Simulation Conference*, pages 706–713, 1995.

[6] Rajive L. Bagrodia and Wen-Toh Liao. Maisie: A Language for the Design of Efficient Discrete-Event Simulations. *IEEE Transactions on Software Engineering*, 20(4):225–238, April 1994.

[7] Jeff S. Steinman. Incremental State Saving in SPEEDES Using C++. In *Proceedings of the 1993 Winter Simulation Conference*, pages 687–696, 1993.

[8] Yoshiko Yasuda et al. Architecture and performance of the Hitachi SR2201 Massively Parallel Processor System. In *Proceedings of the 11th International Parallel Processing Symposium*, pages 233–241, April 1997.

μProfiler: Profiling User-Level Threads in a Shared-Memory Programming Environment

Peter A. Buhr[1] and Robert Denda[2]

[1] University of Waterloo, Waterloo, Ont., Canada
[2] Universität Mannheim, Mannheim, Germany

Abstract. A *profiler* is an important tool for understanding the dynamic behaviour of concurrent programs to locate problems and optimize performance. The best way to improve profiling capabilities and reduce the time to analyze a concurrent program is to use a target-specific profiler that understands the underlying concurrent runtime environment. A profiler for understanding execution of user and kernel level threads is presented, which is target-specific for the μC++ concurrency system. This allows the insertion of hooks into the μC++ data structures and runtime kernel to ensure crucial operations are monitored exactly. Because the profiler is written in μC++ and has an extendible design, it is easy for users to write new metrics and incorporate them into the profiler.

1 Introduction

As programs grow more complex, a greater need arises for understanding their dynamic behaviours, to locate problems and optimize performance. Concurrency increases the complexity of behaviour and introduces additional problems not present in sequential programs. An important tool for locating problems and performance bottlenecks is a *profiler*. However, sequential profiling techniques cannot be trivially extended into the concurrent domain. A concurrent profiler must deal with multiple threads of control, all potentially introducing errors and performance problems. Profiling concurrent programs has been done for performance analysis, algorithm analysis, coverage analysis, tuning, and debugging.

We believe the best way to improve concurrent profiling capabilities and reduce the time to analyze a concurrent program is to use a target-specific profiler that understands the underlying concurrent runtime environment. Our experience in designing several target-specific concurrency tools (high-level concurrent extensions for C++, called μC++ [1], a debugger [2], a profiler [3], and other concurrent toolkits) leads us to conclude that construction of an universal profiler for all languages and concurrency paradigms is doomed to failure.

2 Motivation

The basis of this work is μC++, a shared-memory, user-level thread library running on symmetric multiprocessors (e.g., SUN, DEC, SGI, MP-PC); kernel threads associated with shared memory provide parallelism on multiprocessors, and user threads refine that parallelism. The μC++ environment provides a target-specific debugger for break-point debugging on a user-level thread basis, and experience has shown it aids in the development of robust concurrent programs. Nevertheless, debugging is normally based on a hypothesis concerning the reason for the erroneous behaviour of a program. To reason about general runtime behaviour, including performance analysis, coverage analysis, and tuning, a profiling tool is needed to monitor execution and reveal information at different levels of detail.

Several profilers for concurrent programs exist, but most are general purpose tools with little understanding of the concurrency paradigm. Each concurrent environment provides a different paradigm, which a profiling tool must be aware of in order to provide effective monitoring, analyzing and visualizing of a program's behaviour. The analysis of a concurrent program's performance and algorithmic behaviour becomes more effective and efficient through target-specific profiling, where the profiler has internal knowledge about the runtime system intrinsics and the underlying programming paradigm.

Extendibility is also crucial in designing and implementing an effective profiler, since it is impossible to predict suitable metrics for all imaginable situations. Therefore, a profiling tool should provide a set of general purpose metrics *and* a mechanism enabling a program analyst to quickly develop new problem specific metrics. Hence, an analyst must use knowledge about the profiler, which is easier when the profiler operates as part of the target system and when the metric extensions can be written in a familiar language. Ideally, the same language is used for the profiled program and the profiler extensions.

Finally, a profiler must operate at different levels of detail on concurrent programs to provide the functionality for both exact and statistical profiling. To profile large-scale concurrent programs, selective profiling must be supported: It must be possible to turn profiling on and off dynamically to target specific parts of a large program.

3 Related Work

Most profiling tools have been developed for analyzing the performance of scientific, mostly data-parallel programs, written in a message-based programming environment. For this arena, successful and powerful tools with a wide range of analysis and visualization modules exist. For example, Pablo [4] is a tool with many visual [5] and audio [6] performance data presentation modules. Pablo also introduced a standard trace log format which is adopted by other profile analysis and visualization tools. Another example is Paradyn [7], a tool for profiling large-scale, long-running applications. These program characteristics require some novel instrumentation and analysis methods: dynamic instrumentation insertion and removal based on execution-time profiling information, or user interaction. The results of dynamic instrumentation are promising, but the overhead introduced may reduce effectiveness when profiling code-parallel (in contrast to data-parallel) programs with shorter execution times.

Concurrent profiling tools may be available only as part of the operating system, which allows monitoring of programs, and information about calls for kernel thread creation, synchronization and communication primitives. For example, the Mach Kernel Monitor [8] instruments kernel thread context switching. This approach assumes that the concurrent program's runtime system uses only operating system features, instead of providing portable, user-level thread creation, synchronization and communication primitives.

Among the first profiling tools for a user-level thread-library was Quartz [9]. A target-specific profiling environment for concurrent object-oriented programs is pC++ [10]. pC++ is one of the few cases where the integrated performance analysis environment TAU [11] was implemented in concert with the language and runtime system. However, the design of pC++/TAU incorporates most of its profiling functionality into the preprocessor and runtime system, so extending the profiling metrics by the program analyst takes more effort. The tight coupling between the language/runtime system and the profiling tool makes integration into other existing thread-libraries infeasible.

4 μProfiler

μProfiler [3] is a concurrent profiler, running on UNIX based symmetric shared-memory multiprocessors, that achieves our goal of target-specific, extendible, fine-grained profiling on a user-level thread basis. μProfiler supports the μC++ shared-memory programming model, which shares all data and has multiple kernel and user-level threads. Profiling μC++ programs requires incorporating both concurrent and object-oriented aspects, i.e., profiling different threads of control at the per-object level.

Profiling sequential programs is non-trivial but well-understood. Additional challenges arise when profiling concurrent programs in a shared-memory environment similar to μC++. Since the environment provides user-level tasks, the profiler must monitor the program's activity at that level. The profiler also needs internal knowledge about the runtime system to identify and monitor each executing task independently and exactly. The μProfiler design deals with these challenges and presents a mechanism to effectively integrate extendible profiling into the concurrency system.

μProfiler is a concurrent program written in μC++, executing concurrently with the profiled μC++ application (see Fig. 1). A *cluster*, which groups user and kernel threads and restricts the execution of those user threads by the kernel threads in the cluster, is the μC++ capability which enables concurrent application and profiler execution. The user threads in the profiler cluster monitor execution of the runtime kernel and other clusters using direct memory reads via shared memory. On multiprocessor computers, the kernel thread in the profiler cluster executes the profiler user threads in parallel with the application. If the amount of the monitoring is large, more kernel threads can be added to the profiler cluster to increase parallelism. So application performance is degraded only by the contention created by profiler operations. This cost can be 100 to

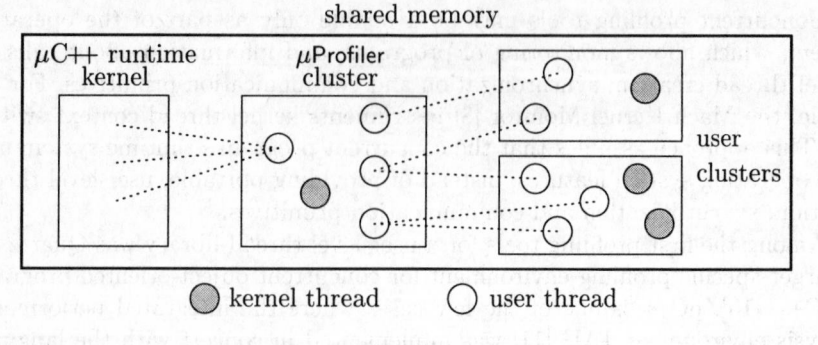

shared memory

Fig. 1. Integration of μProfiler into the profiled program

1000 times less than monitoring from a separate UNIX process, which requires cross-address-space reads. More complex monitoring is thus possible, while still having only a small effect on the application.

To access μProfiler, compilation flags -profile and -kernelprofile cause the necessary instrumentation insertion and linking with the profiling libraries. The first flag profiles only the user program; the second flag profiles the μC++ kernel calls made by the program. (The latter information is often inappropriate and confusing to users.) When the program starts, a menu appears, from which a user selects several builtin metrics, after which the program is run, and the metric output appears. Thus, in the simplest case, instrumentation insertion and activation of the profiling modules is completely transparent to the programmer. Additionally, parts of a program may be compiled with or without the profile flag(s), and then linked together, creating an executable where selected parts are instrumented and profiled. Finally, even more precise control is available through routines in the μC++ runtime system to turn profiling on and off for a particular thread at any point during execution.

Because μProfiler is truly integrated with μC++, it was possible to insert hooks into the μC++ runtime kernel to ensure crucial operations are monitored exactly, such as user and kernel thread creation/destruction, and migration of user and kernel threads among clusters. Purely statistical monitoring and dynamic instrumentation could miss some of these events. Also, dynamic instrumentation is considered too expensive when profiling programs with short or intermediate execution times. Exact routine counts are obtained via static instrumentation insertion at compile-time using shared trampolines. The C compiler -pg option is used to generate routine-entry instrumentation; the routine-entry instrumentation is augmented to generate routine-exit instrumentation. In addition, arbitrary hooks can be inserted into user code by the program analyst.

All hooks can be dynamically activated and deactivated on a per thread basis, but only when the profiler is present in the application (i.e., the existence of the profiler is checked for dynamically inside the runtime kernel). Each activated hook results in the profiled thread sending event(s) to the profiler, which passes the information to the active profiling monitors. Figure 2(a) shows a (simple)

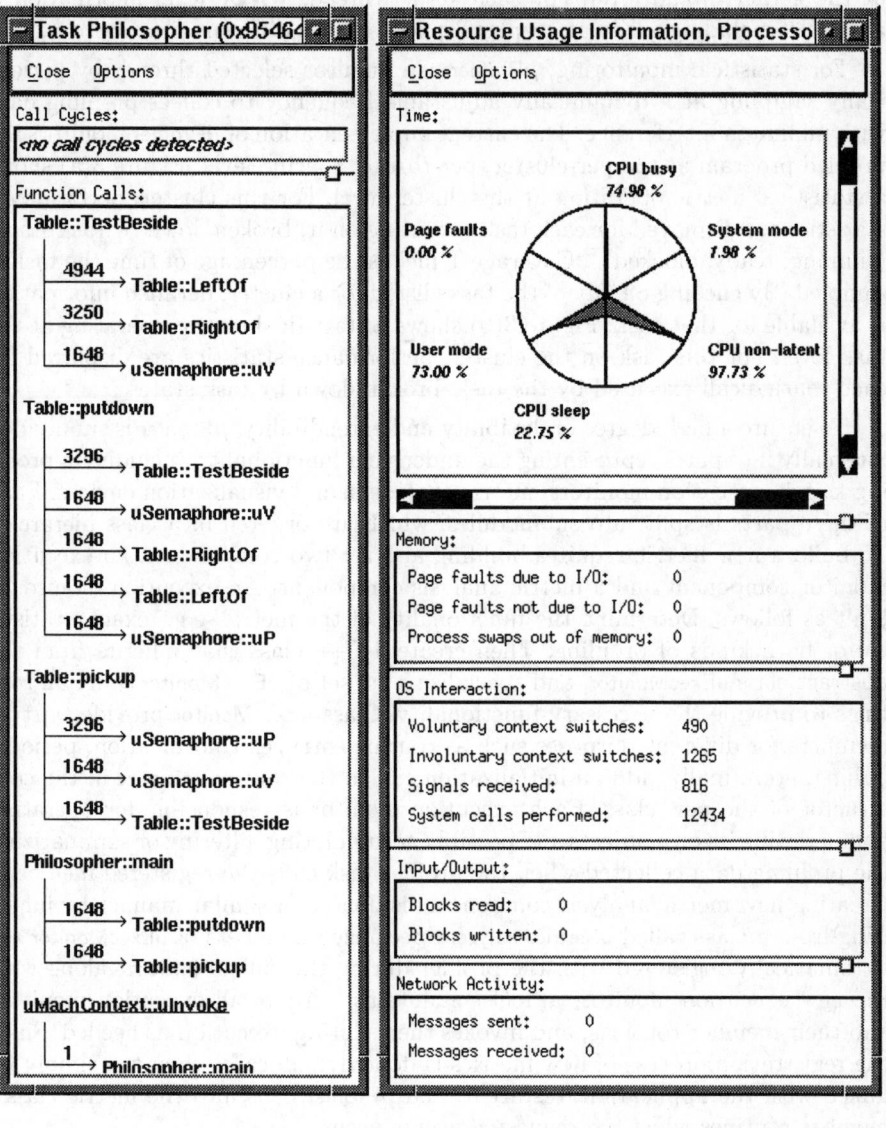

(a) Exact User Thread Metric   (b) Exact Kernel Thread Metric

Fig. 2. μProfiler Exact metrics

exact metric operating at the user thread level. For each user thread, *gprof*-like [12] routine call information is available, including call cycles. Function calls from within each executed routine are presented with the corresponding routine call count information. Figure 2(b) shows a (simple) exact metric operating at the kernel thread level. For each kernel thread, *UNIX* level information is available, including a Kiviat graph to quickly relate different time metrics.

For statistical monitoring, µProfiler can monitor selected threads by periodically sampling at a dynamically adjustable frequency to collect profiling data with minimum interference. The current implementation of µProfiler monitors the profiled program at the per-cluster, per-thread, routine level. Figure 3(a) shows a statistical metric operating at the cluster level. For this cluster, performance statistics are displayed for each task executing on it, broken down by task states (running, ready, blocked). "Coverage Time" is the percentage of time the task is sampled. By clicking on any of the tasks listed for a cluster, detailed information is available for that task. Figure 3(b) shows a statistical metric operating at the task level. For this task on the cluster, performance statistics are displayed for each routine call executed by the task, broken down by task states.

To ensure a high degree of flexibility and extendibility, µProfiler is subdivided internally into parts representing the underlying functionality, including a profiling kernel, execution monitors, metric analyzers, and visualization devices. Each of these parts is split into submodules, which are ordered in a class hierarchy. To build a new metric requires building at least two components: an execution monitor component and a metric analysis component. An execution monitor is built as follows. Determine the functionality of the metric: e.g., exact, statistical or both kinds of profiling. Then create a C++ class that inherits from the abstract class uExecMonitor, and specialize a subset of uExecMonitor's virtual routines to provide the necessary functionality. Class uExecMonitor provides virtual members for different purposes such as routine entry/exit notification, periodic polling, etc. Finally, add an initialization call to the routine Initialize in the constructor of the new class. Each execution monitor is responsible for operating and updating its own objects and possibly accumulating, filtering or summarizing the profiling data collected when the profiler task calls the registered members. Creating new metric analysis components is done in a similar manner by inheriting from a class called uMetricAnalyze. Specialized members of uExecMonitor are automatically registered with the profiler during the call to Initialize along with the new execution monitor. µProfiler maintains a list of all execution monitors and their member routines, and invokes them during execution as needed. Since the registration process of new metrics is done dynamically, they can simply be linked with the application, restart it, and µProfiler calls into the metric class's member routines when the requested events occur.

Additional reuse is provided by inheriting from existing metrics that come with the µProfiler library or previously built by the program analyst. All µProfiler metrics conform to the above mechanism. uSPMonitor, for instance, is an execution monitor that statistically samples a task and measures the time the task spends on a certain cluster in a certain routine in a particular state. Because uSPMonitor is based on statistical sampling, it inherits from uExecMonitor and specializes the poll routine, in which the data collection is performed.

Through this mechanism, an analyst can efficiently extend the functionality of µProfiler by metrics that fit the analyzed problem much better than general

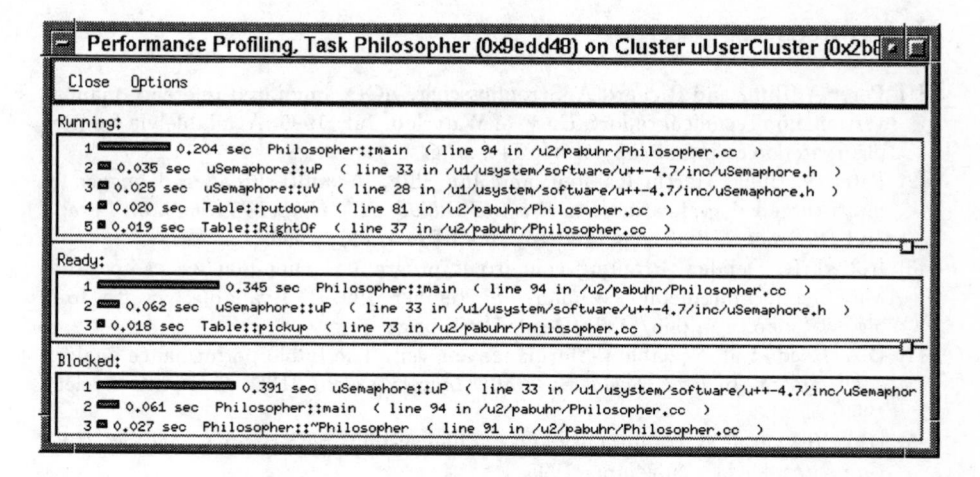

(a) Cluster Performance Metric

(b) Task Performance Metric

Fig. 3. μProfiler performance metric

purpose metrics created by the developers, resulting in profiling results that directly correspond to the problem under investigation. This approach enables an analyst to extend μProfiler's functionality with any metric, analysis or visualization device using exact or statistical monitoring. It is more important to integrate the basic functionality for different execution monitoring, analyzing and visualizing methodologies on which both general and problem-specific modules can operate, than to build a fixed set of highly sophisticated metrics.

Concurrency is also part of some object-oriented languages, e.g., μC++. μProfiler can identify corresponding objects (both caller and callee side) when a monitored task invokes an object's member routines. While there are no predefined metrics using this feature, we anticipate them soon.

5 Conclusion

Concurrent systems have complex dynamic behaviour with significant implicit information embedded in the runtime environment. Our claim is that target-specific profilers can do a better job extracting and displaying information from this environment. We show that tight integration possible with a target-specific profiler, i.e., between μProfiler and μC++, results in better information gathering at lower cost, and the ability to easily add new metrics through a single programming language. The μProfiler displays are simple but informative, requiring the analyst to manually locate performance issues, e.g., hot spots, by examining the data. We have found manual determination to be straightforward, and have discovered several performance problems using μProfiler while examining both μC++ and μC++ applications to understand their dynamic behaviour.

References

[1] Peter A. Buhr and Richard A. Stroobosscher. μC++ annotated reference manual, version 4.6. Technical report, Univ. of Waterloo, July 1996. Available via ftp from plg.uwaterloo.ca in pub/uSystem/uC++.ps.gz.

[2] Peter A. Buhr, Martin Karsten, and Jun Shih. A multi-threaded debugger for multi-threaded applications. In *Proc. SPDT'96: SIGMETRICS Symp. on Parallel and Distributed Tools*, pages 80–87. ACM Press, May 1996.

[3] Robert R. Denda. Profiling concurrent programs. Diplomarbeit, Universität Mannheim, Mannheim, Germany, September 1997. Available via ftp from plg.uwaterloo.ca in pub/MVD/DendaThesis.ps.gz.

[4] D.A. Reed et al. Scalable performance analysis: The Pablo performance analysis environment. In *Proc. Scalable Parallel Libraries Conf*. IEEE Computer Society, 1993.

[5] D.A. Reed et al. Virtual reality and parallel systems performance analysis. *IEEE Computer*, 28(11), November 1995.

[6] Tara Maja Madhyastha. A portable system for data sonification. Master's thesis, Rutgers State Univ., 1990.

[7] B. Miller et al. The Paradyn parallel performance measurement tools. *IEEE Computer*, 28(11), November 1995.

[8] T. Lehr et al. MKM: Mach kernel monitor description, examples and measurements. Technical report, Carnegie-Mellon Univ., March 1989. PA-CS-89-131.

[9] T.E. Anderson and E.D. Lazowska. Quartz: A tool for tuning parallel program performance. In *Proc. 1990 SIGMETRICS Conf. on Measurement and Modeling of Computer Systems*, pages 115–125, Boston, May 1990.

[10] A. Malony et al. Performance analysis of pC++: A portable data-parallel programming system for scalable parallel computers. In *Proc. 8th Int. Parallel Processing Symp.*, pages 75–85, Cancun, Mexico, April 1994.

[11] D. Brown et al. Program analysis environments for parallel language systems: The τ environment. In *Proc. 2nd Workshop on Environments and Tools For Parallel Scientific Computing*, pages 162–171, Townsend, Tennessee, May 1994.

[12] S. L. Graham, P. B. Kessler, and M. K. McKusick. GPROF: a call graph execution profiler. *SIGPLAN Notices*, 17(6):120–126, June 1982.

Evaluating a Multithreaded Runtime System for Concurrent Object-Oriented Languages

Antonio J. Nebro, Ernesto Pimentel and José M. Troya

Universidad de Málaga, Málaga, Spain

Abstract. Traditionally, the use of multithreading capabilities of operating systems has been considered inadequate for implementing concurrent object-oriented languages because of their inefficiency and non-portability. However, current operating systems encourage programmers to use threads to manage concurrent activities, since they offer a number of advantages such as multiprocessing capabilities and thread communication through shared memory. To explore these issues, we have developed Lince, a multithreaded runtime system for concurrent objects. We describe Lince and its design philosophy and analyze its performance. The use of popular threads packages allows us to simplify system design and enhance portability. The overhead of using threads for implementing concurrent objects is negligible for medium and coarse grain applications, although it can be too expensive for those requiring many fine-grained objects.

1 Introduction

Concurrent object-oriented programming is a paradigm that tries to take advantage of object orientation in the development of concurrent software. Under this model, parallel programs are collections of concurrent objects that interact by invoking the operations they define in their interfaces. One of the challenges of concurrent-object oriented computing is to implement this high-level programming model efficiently and in a portable fashion. The term portable means not only that parallel programs can run on computers with different processors and/or operating systems, but also in different parallel architectures. Nowadays, the most popular parallel architectures are multiprocessors and networks of workstations, so a concurrent object-oriented program should take advantage of these kinds of systems.

Traditionally, one of the drawbacks of concurrent object-oriented languages has been their inefficiency. However, recent work in this field reveals that this assertion is not necessarily true. The design of efficient runtime systems [1], sometimes combined with aggressive compilation techniques [2], has shown that concurrent object-oriented languages developed on top of such systems can attain sequential efficiency. Our approach is that current operating systems offer a set of facilities that can allow us to build efficient and portable runtime systems for concurrent object-oriented languages. In particular, we focus on multithreading. Modern operating systems encourage programmers to use threads to manage concurrent activities because they offer a number of advantages, such as computation and I/O overlapping, multiprocessing capabilities, and thread communication through shared memory. Our proposal is based on the application of distributed shared memory techniques (DSM) for dealing with distribution issues [3, 4] and the use of threads for implementing concurrent objects. The result is a runtime system which we call Lince. In this paper we describe the Lince system and measure its performance in several mono and multiprocessor systems. In particular, we focus on the analysis of the influence of multithreading on the cost of basic operations, such as object creation and method invocation, in the case of local objects. Lince's performance in distributed systems is beyond the scope of this paper.

A key point of this work is that multithreading can lead to simpler and portable runtime systems, giving over to the operating system some functions (e.g., object scheduling) at the cost of an overhead that could be acceptable for a wide number of applications. Although a number of threads API exist, the increasing availability of Pthreads and the fact that other popular threads packages (Solaris and Microsoft's Win32 threads) share a common subset of basic features [5], allows us to consider threads as a choice to be taken into account. The basic design objectives of Lince are to: a) provide a platform to write concurrent object-oriented programs, but offering services that can be used by compilers; b) facilitate the portability of parallel applications to a broad range of platforms, and; c) achieve good performance.

The rest of the paper is organized as follows. In Sect. 2, we discuss the object model assumed in this work. The architecture of the Lince system is described in Sect. 3. The next section covers implementation details and the evaluation of the system. Related work is discussed in Sect. 5. Finally, Sect. 6 presents some conclusions and discussion of future work.

2 Concurrent Objects in Lince

Our work has to take place in the context of an object model that defines the structure of objects and their behavior. We have chosen a generic and simplified object model. We define a concurrent object as an entity that has an internal state composed of a set of hidden variables, and a public interface composed of a set of operations. State variables can only be accessed through operation invocation. Object activities are reduced to invoking other object operations, creating additional objects and modifying its internal state because an operation has been performed. Examples of languages that fit into this basic object model are those based on the actor model [6], like ABCL [7] and HAL [8].

The proposed object model extends the basic asynchronous operation invocation scheme of the actor model also allowing synchronous invocations. We distinguish two kinds of operations: commands and queries. A command is an operation that modifies the internal state of the invoked object, but without returning information about it. A query is an operation that returns information about the object, but without modifying its internal state. Commands and

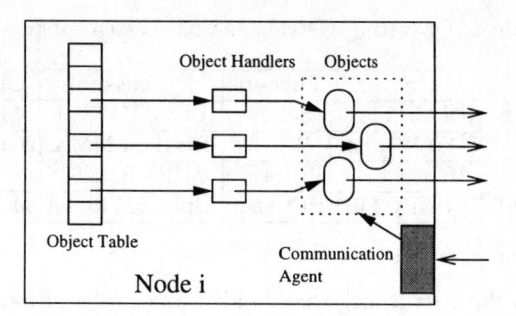

Fig. 1. Architecture of the implementation

queries present different behavior: commands are asynchronous operations, while queries are synchronous operations. According to this distinction, objects synchronize when a query is invoked, according to a wait-by-necessity mechanism [9]. This concurrency model fits into Meyer's proposal of integrating concurrency and object-orientation [10].

The Lince system relies on multithreading capabilities supplied by the operating system to tackle object management and communications in the case of local objects. We believe, unlike some previous works [11], that current thread packages can provide a portable and efficient abstract layer. Modern operating system threads support reduces the cost of multithreading while offering additional advantages such as exploiting multiprocessing in a transparent way. Features like memory sharing enables the use of aggressive compiler techniques such as inlining or speculative optimizations.

The application of DSM techniques in Lince allows us to modify object location dynamically by replicating or migrating objects. The Lince implementation scheme adopts the entry memory consistency model [12], that reduces communication latency while requiring a smaller number of messages than other consistency models. The drawback is that method invocations must be enclosed between a pair of acquire and release operations on a lock that is associated to the replicated object. Acquire operations can be exclusive or shared, so objects are accessed according to a multiple-reader/single-writer scheme. To avoid increasing the complexity of the programming model, this invocation scheme is applied to all the objects, replicated or not.

3   Lince Multithreaded Architecture

The architecture of the current implementation is shown in Fig. 1. In each node there is a process that includes the runtime system and the concurrent objects. Objects are implemented using one thread per object, and each thread execute the same scheduling function. The runtime system is composed of the object table, the communication agent (CA), and a set of structures, like the node identifier and the host table. Each entry in the object table is a pointer to an

Table 1. Operating systems and hardware configurations

Computer	OS	Threads	Processor)	Clock	Memory
Sun Ultra 1	Solaris 2.5	Solaris	UltraSPARC I	143MHz	64MB
PC	NT 4.0 Workst.	Win32	Pentium MMX	200MHz	64MB
SGI O2K	IRIX 6.4	Pthreads	MIPS R10000	196MHz	4GB
AlphaSvr 4100	DECUNIX 4.0D	Pthreads	Alpha 21140-AA	300MHz	256MB

object handler, a data structure that holds object related data such as: thread identifier, the incoming message queue, a lock for ensuring mutual exclusion, and a pointer to a reply box. The CA is an entity whose main aim is to receive messages from objects in the other nodes and deliver them to the target objects. When objects are local they communicate without using the CA, by putting the messages directly into the invoked object's queue.

The Lince system provides a set of services that are used through object handlers. These services are classified in object management (basic services for object creation and operation invocation), object information (a set of calls that report information about an object, like its locality or its state), and optimized services. These last ones are optimized calls that replace some of the basic services. For example, when only an isolated command or query is going to be invoked, instead of issuing three operations (acquire + command/query + release), the system provides services such as `CommandAcquire` or `QueryAcquire`, that produce the same effect but only requiring two messages.

Although the optimized services improve the performance of the general invoking scheme by requiring fewer messages, the communication costs are still far from the objective of sequential efficiency when objects are local. Nevertheless, the use of threads for implementing concurrent objects allows the employment of techniques such as inlining. Thus, an operation invocation can be replaced by a simpler piece of code, like the code of the operation (the function call and return are eliminated) or even the state variables of the object.

4 Implementation Details and Evaluation

Our current implementation of the Lince system is a prototype written in C++ [13]. We have used three different thread packages: Pthreads [14], Solaris threads [15] and Win32 threads [16]. At present, there is a port to the following operating systems: Solaris 2.X, Digital Unix 4.0, IRIX 6.4, and Windows NT/95. We evaluated the system on the platforms shown in Table 1.

4.1 Basic Services Evaluation

In this section we show the costs of basic operations, such as object creation and operation invocation.

Table 2 shows the cost of object creation in these systems. This process includes obtaining a new object handler and invoking the `CreateObject` service.

Table 2. Costs of basic operations (in microseconds)

	AlphaServer 4100	Origin-2000	Ultra 1	PC
Object creation	970	1600	4300	1600
Acquire+Query+Release	52	42	234	130
QueryAcquire	46	30	160	80
Query (already acquired)	0.8	1	1	10

The times obtained depend directly on the primitives provided for thread creation, and are about three orders of magnitude worse than those obtained in
other works [1]. These results indicate that the thread approach is not suitable
for intensive fine-grained object-oriented computing, but they could be considered acceptable if most of the fine-grained objects in the application are long-
lived. We also show the cost of invoking a query operation on a local concurrent
integer object using the acquire-query-release scheme and the optimized service
QueryAcquire, and the cost of invoking a query when the object is already ac-
quired. These results indicate that in the worst case the cost of a local invocation
is in the order of tens of microseconds.

4.2 Analysis of a Matrix Multiply Program

We have coded a parallel matrix multiply program to measure the impact of
increasing the number of objects while decreasing the grain of computations they
have to perform. The parallel algorithm is based on dividing the result matrix
into submatrices that can be computed in parallel by multiplier objects. For the
sake of simplicity, the program multiplies square matrices of floats, and the result
matrix is divide into 4^N submatrices. Thus, by varying N from 0 to 4, the same
computation will be carried out by a single process that contains between 1 and
256 multiplier objects. As the matrix objects to be multiplied receive mostly read
operations, methods for accessing matrix elements are inlined. Thus, multiplier
objects can directly access the internal state of the matrix objects.

Baseline results for the matrix multiply program executions, multiplying two
256 and two 1024 matrices, using one processor, appear in Table 3. The times
include the creation of the multiplier objects, operation invocations and termination detection.

At first glance, the results of multiplying the 1024 matrices are a bit surprising, because the times tend to decrease when increasing the number of threads,
and in each execution the number of computations remains constant and we
have to add the overhead of using threads. It is difficult to explain this behavior, but we must consider several issues in order to draw a conclusion. First,
sequential matrix multiplication is a processor-bound process which has been
traditionally penalized by the scheduling policy of UNIX systems (and Windows
NT). Therefore, it is possible that the multithreaded version behaves better in
terms of scheduling. Second, taking into account that the executions take tens
of seconds, the cost of creating up to 256 objects is negligible. This is not the

Table 3. Times obtained multiplying two 256 and two 1024 square float matrices using one processor (in seconds)

	256 x 256				1024 x 1024			
	AlphaServer	Origin	Ultra 1	PC	AlphaServer	Origin	Ultra 1	PC
Sequential	0.73	0.51	1.54	1.15	49.5	35.4	117.8	85.7
1 Mult.	0.78	0.34	1.32	1.12	59.5	25.7	98.2	81.4
4 Mults.	0.77	0.41	1.32	1.12	59.6	25.6	97.8	81.4
16 Mults.	0.78	0.55	1.41	1.13	59.5	23.6	97.0	81.4
64 Mults.	0.82	2.82	1.48	1.21	59.5	22.8	95.2	78.7
256 Mults.	1.02	1.14	2.16	1.51	59.6	23.0	90.5	73.5

case of the 256 matrix multiplication, where increasing the number of threads penalizes slightly performance up to 16 multipliers. When using a higher number of multipliers performance tends to degrade rapidly.

We repeated the same experiments in the multiprocessor systems using all the available processors (Table 4). Speedups were obtained by dividing the time of the execution with one multiplier object. We observed that in the Alphaserver the speedups were very similar and close to linear in the 1024 matrix multiplication, but they decrease in the 256 matrix experiments. In the case of the Origin-2000, the results are not very good. A reason can be related to the scheduling policy of the IRIX operating system, which seems to create the new threads in the same node that the creating thread and then it balances the load dinamycally. As the grain of computations of the 256 matrix multiplication is small, the most of computations will be executed probably using fewer processors than available.

5   Related Work

This section reviews previous works related to run-time support for implementing concurrent objects. The Illinois Concert system combines aggressive compiler and runtime techniques for implementing fine-grained concurrent object-oriented programs in sequential and parallel systems [2]. Concert has demonstrated sequential performance in several benchmarks [17]. Our work differs mainly in that, to achieve portability and take advantage of multiprocessing, Lince is implemented using OS threads.

StackThreads is a runtime system that has been used to implement ABCL, one of the first concurrent-OO programming languages [1, 18]. Like Concert, the goals of StackThreads are in obtaining sequential efficiency, but focusing on runtime techniques and not on compiling analysis and optimization. No support for object migration or replication is provided, nor are they considered in the ABCL language.

Several projects focus on portable runtime support for object-oriented languages based on C++, like Mentat and CHARM++. Mentat is an object-oriented parallel system based on a dataflow computation model. Its runtime system is implemented using processes [19], so Mentat is mainly suitable for

Table 4. Times (in seconds) and speedups obtained multiplying two 256 and two 1024 square float matrices in the multiprocessor systems

	256 x 256				1024 x 1024			
	AlphaServer 4100 (4 processors)		Origin-2000 (16 processors)		AlphaServer 4100 (4 processors)		Origin-2000 (16 processors)	
	Time	Speedup	Time	Speedup	Time	Speedup	Time	Speedup
1 Mult.	0.76	1	0.34	1	59.7	1	23.7	1
4 Mults.	0.20	3.8	0.27	1.25	15.2	3.93	7.4	3.20
16 Mults.	0.22	3.4	0.15	2.42	15.4	3.87	3.0	7.90
64 Mults.	0.24	3.2	0.16	2.12	15.2	3.93	2.2	10.77
256 Mults.	0.47	1.5	0.59	0.57	15.2	3.93	2.4	9.87

medium-to-coarse-grain applications. CHARM++ [11] is a C++ extension that classifies objects into sequential, concurrent, replicated, shared, and communication objects. In Lince we adopted a more uniform object model, where all the objects are concurrent, shared, and can be replicated.

6 Conclusions and Future Work

We have described Lince, a multithreaded runtime system for implementing concurrent object-oriented languages. Implementing concurrent objects using threads provided by modern operating systems allows simplifying system design and taking advantage of multiprocessing capabilities. Portability is enhanced because most thread packages offer a common subset of features. The system is based on an object model that is suitable for replicating objects, so the programming model imposes several requirements, such as explicitly acquiring and releasing objects, and the distinction between command and query operations.

Lince performance measurements show that creating objects is expensive, while the overhead of method invocation is about several tens of microseconds in the worst case. Results of a parallel matrix multiply program reveal that increasing the number of threads while decreasing granularity can lead to better performance, contrary to expectations.

We conclude that the overhead of using threads for implementing concurrent objects can be acceptable for medium and coarse applications, although it is too expensive for those requiring many fine-grained objects. Tuning the Lince runtime system for better performance and building representative application workloads are topics of future work.

Acknowledgments E. Pimentel was supported in part by the "Comisión Interministerial de Ciencia y Tecnología" (CICYT) under grant TIC95-0433-C03-02.

References

[1] K. Taura, S. Matsuoka, and A. Yonezawa. An efficient implementation scheme of concurrent oriented-languages on stock multicomputers. In *4th ACM SIGPLAN Symposium on Principles and Practice of Parallel Programming*, May 1993.

[2] Andrew A. Chien, V. Karamcheti, and J. Plevyak. The Concert system—compiler and runtime support for efficient fine-grained concurrent object-oriented programs. Technical report, Department of Computer Science, University of Illinois, Urbana, Illinois, June 1993.

[3] Antonio J. Nebro, Ernesto Pimentel, and José. M. Troya. Integrating an entry consistency memory model and concurrent object-oriented programming. In Christian Lengauer, Martin Griebl, and Sergei Gorlatch, editors, *Third International Euro-Par Conference*, volume 1300 of *Lecture Notes in Computer Science*. Springer-Verlag, August 1997. Passau, Germany.

[4] Antonio J. Nebro, Ernesto Pimentel, and J. M. Troya. Distributed objects: An approach based on replication and migration. Accepted to be published in the Journal of Object Oriented Programming (JOOP), 1998.

[5] Sun Microsystems, Inc. *Multithreaded Implementations and Comparison: A White Paper*, 1996.

[6] Gul Agha. *ACTORS: A Model of Concurrent Computation in Distributed Systems*. The MIT Press, 1986.

[7] A. Yonezawa. *ABCL: An Object-Oriented Concurrent System*. MIT Press, 1990.

[8] C. Houck and G. Agha. Hal: A high level actor language and its distributed implementation. In *21st. International Conference on Parallel Processing*, August 1992. St. Charles, IL.

[9] Denis Caromel. Towards a method of object-oriented concurrent programming. *Communications of the ACM*, 36(9), September 1993.

[10] Bertrand Meyer. Systematic concurrent object-oriented programming. *Communications of the ACM*, 36(9), September 1993.

[11] L. V. Kale and S. Krishnan. Charm++: A portable concurrent object-oriented language. In *OOPSLA'93*, 1993.

[12] Brian B Bershad and M J Zekauskas. Midway: Shared memory parallel programming with entry consistency for distributed memory multiprocessors. Technical Report CMU-CS-91-70-170, Carnegie-Mellon University, 1991.

[13] B. Stroustrup. *The C++ Programming Language. Second Edition*. Addison-Wesley, Summer 1991.

[14] D. R. Butenhof. *Programming with POSIX Threads*. Addison-Wesley, 1997.

[15] Sun Microsystems, Inc. *Multithreaded Programming Guide. Solaris 2.6*, 1997.

[16] T. Q. Phan and P. K. Garg. *Multithreaded Programming with Windows NT*. Prentice-Hall, 1996.

[17] J. Plevyak, X. Zhang, and A. A. Chien. Obtaining sequential efficiency for concurrent object-oriented languages. In *22nd Symposium on Principles of Programming Languages (POPL'95)*, 1995. San Francisco, California.

[18] K. Taura and A. Yonezawa. Fine-grain multithreading with minimal compiler support—a cost effective approach to implementing efficient multithreading languages. In *ACM SIGPLAN'97 Conference on Programming Language Design and Implementation (PLDI)*, June 1997. Las Vegas, Nevada.

[19] A. S. Grimshaw, J. B. Weissman, and W. T. Strayer. Portable run-time support for dynamic object-oriented parallel processing. *ACM Transactions on Computer Systems*, 14(2), May 1996.

Object-Oriented Run-Time Support for Data-Parallel Applications*

Hua Bi, Matthias Kessler, and Matthias Wilhelmi

GMD Institute for Computer Architecture and Software Technology, Berlin, Germany

Abstract. We present a C++ template run-time library, PROMOTER, and discuss run-time support for data-parallel applications. The PROMOTER run-time library provides a uniform framework for data-parallel applications, covering a broad spectrum of granularity, regularity and dynamicity. It supports user-defined data structures ranging from dense to sparse arrays, regular to irregular index structures and data distributions. The object-oriented design and implementation of the PROMOTER run-time library not only provides an easy data-parallel programming environment, but also leads to an efficient implementation of data-parallel applications through object reuse and object specialization.

1   Introduction

A frequently used model for developing parallel applications on distributed-memory multiprocessors is the data-parallel programming model in which parallelism is achieved by partitioning large data sets between processors, and having each processor work only on its local data.

In this paper we discuss the run-time support for data-parallel applications provided by the PROMOTER run-time library (PRL). By the object-oriented design principle, the PRL provides a uniform interface to support data-parallel applications on both dense and sparse arrays or data structures. By sparse arrays we mean not only regular ones but also irregular ones. A regular sparse array may have an index set with a regular scheme, like a band matrix, while in an irregular sparse array indices are generated irregularly at run-time. A uniform interface is necessary because in many scientific applications, operations on both dense and sparse arrays are coexisting.

In the following, we first present a run-time model for data distribution in which data distribution descriptors as run-time support are introduced. Then we discuss how to use this run-time support in computation and communication. Finally, we analyze some performance results, compare our approach with related works, and give the concluding remarks.

2   Data Distribution

Our approach assumes that a set of (virtual) SPMD processes runs in parallel on a distributed-memory multiprocessor. Each process has at least one control flow

---

* This work is supported by the Real World Computing Partnership, Japan.

(possibly multi-threaded) and its own address space. In our terms, the address space of a process is called a domain. Each process can only access elements on its own domain, and can communicate with other processes by receiving and sending elements from and to other domains.

2.1 Data Distribution Descriptor

Most scientific applications are using dense and/or sparse arrays for modeling their data structures. Arrays are always indexed by so-called *subscripts*. The space containing all subscripts is called *index space* for that array. An index space describes the spatial structure of an array.

In a parallel implementation, because of data distribution (the partitioning of data between processors) there are two kinds of index spaces. A global index space with global subscripts describes the spatial structure of an array in a sequential context, while a local index space with local subscripts describes the spatial structure of an array in a parallel context (that is, after applying data distribution). In our work, the global index space should be provided by users, because it is problem-oriented. The local index space and its relation to the global index space are maintained by the PRL in so-called *data distribution descriptors*. They are defined as *Distribution* classes with the following structure:

- a *map* method that returns a domain number for any global subscript
- a *transform* method that returns a local subscript for any global subscript belonging to the local domain
- an *Iterator* class to iterate over all indices that belong to the local domain, and from which the global and local subscripts of the *current* index can be retrieved
- an *Allocator* super-class that provides information for the allocation of local memory.

A *map* method is normally implemented by an arithmetic function. In this way, the PRL does not restrict mapping strategies to some pre-defined ones such as block-cyclic mapping.

The *transform* method and the *Iterator* class can be simply implemented for (regular) dense arrays. For (irregular) sparse arrays, they have to be implemented by searching a table to get results in the worst case. In our approach, we only require the local information from the *transform* method and the *Iterator* class. In other words, the data distribution descriptor itself is distributed. In this way, the worst case invokes a search only restricted to the local index space and its performance related to memory size and search time is scalable.

An *Allocator* class determines how the local data set should be allocated. PRL provides different *Allocator* classes to allocate local data in the form of *vector*, *matrix*, *binary tree* or *hash table*.

The current PRL provides a broad spectrum of *Distribution* classes for dense arrays and sparse arrays with different mapping strategies (see Fig. 1).

Users or compilers can select a suitable distribution class according to the problem to be solved. One of the most distinguished features of the PRL is that

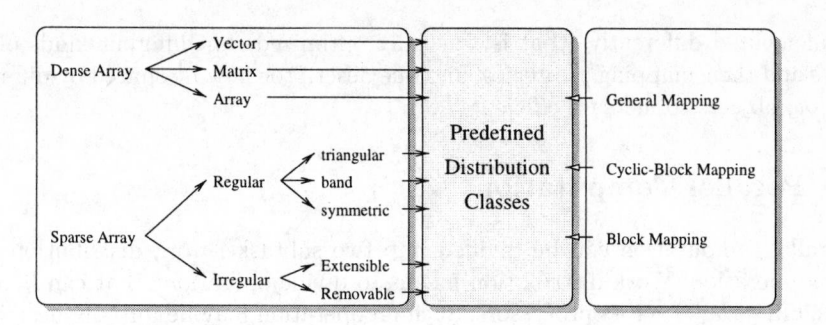

Fig. 1. Distribution Classes

all data distribution descriptors in it have a uniform type interface. The PRL is realized by class and function templates that are parameterized over data types such as *Distribution*. Different data distribution descriptors, with respect to different natures of arrays and their mapping strategies, can be implemented and then used in the PRL. They only have to follow the type interface defined by the PRL. So it is open for users or compilers to provide their own implementation of *Distribution* and other classes.

The *Distribution* classes can be implemented as STL adaptors. The STL container and iterator classes cannot be directly used like the *Distribution* classes because the problem of data distribution should be dealt with explicitly.

2.2 Distributed Object

A distributed object is a collection of data elements which are partitioned and then allocated on local memory of a distributed-memory multiprocessor.

```
template<class DT, class ET> Disobj {
public:
        Disobj(const DT& t);
        ~Disobj();
        ET& operator[](const int index) const;
};
```

A distributed object is defined as a class template over a *Distribution* class DT and an element type ET, with the meaning that the data elements are of the type ET and are partitioned and allocated according to the *Distribution* class DT. The operator [ ] provides access to local data elements through a local subscript.

With a uniform interface to data distribution descriptors, *Disobj* classes can be implemented as generic class templates. In fact, the implementation of *Disobj* classes differs only on different *Allocator* classes. That is, the PRL provides the corresponding *Disobj* class templates with respect to different *Allocator* classes.

Data distribution is supported by providing descriptors and carriers respectively. We can benefit from this decomposition in two aspects: descriptors can be implemented differently (that is, specially optimized) for different kinds of arrays and their mapping strategies, and one descriptor may be shared by different *Disobj* objects to increase efficiency.

3 Parallel Computation

Parallel computation can be divided into two subtasks: work distribution and local operation. Work distribution means to divide operations that can be executed in parallel across processors. Such an operation may be initially described by a loop over global subscripts in a global index space. Work distribution is performed in two steps. In the first step, the loop is divided into a set of local loops with respect to the local domain. In the second step, all global subscripts involved are transformed to local subscripts.

A loop over a global index space can be transformed to a loop over a local index space by calling the iterator of the *Distribution* class of a target *Disobj* object. At each iteration, the current subscripts (both global and local) are obtained. The global subscript is then used in a test within the loop to filter unnecessary operations as defined by the original loop, and the local subscript is used to access the target data element. For all source *Disobj* objects, each global subscript is computed as a function of the current global subscript of the target element. Then the corresponding local subscript is found by calling the *transform* method in a *Distribution* class. Non-distributed objects can also be used in parallel computation, as they are duplicated at each domain.

The above implementation can be further improved by the following two optimizations: First, since the work distribution described above is only dependent on data distribution descriptors, it is possible to collect all local subscripts for an operation in the beginning and then reuse them multiple times, e.g., within a loop. Second, because in the collection of local subscripts, we only use an *Iterator* class from a *Distribution* class. It is possible to define some special *Iterator* classes which only iterate over a subset of local indices with respect to a *Distribution* class, for example, a row, a column, or a diagonal from a matrix.

4 Communication

Communication in data-parallel applications can be divided into two kinds: point-to-point and collective. Based on the run-time support provided by a *Distribution* class, the point-to-point communication can easily be implemented. Each process checks if it is the owner of a source element or a target element by calling the *mapping* method. If a process is the owner of source or target elements, a local subscript for that element is found by calling the method *transform*, and then send or receive routines are called to perform the communication.

The collective communication is performed in the so-called *inspector/executor* paradigm[1]. First, an *inspector* routine is called to build a communication schedule that describes the required data motion, and then an *executor* routine is called to perform the data motion (sends and receives). The communication scheduling phase must be performed at run-time for dense arrays, if some parameters (rank of arrays, loop parameters, or the number of available processors) or data accesses are unknown at compile-time. For sparse arrays, even the spatial structure is unknown at compile-time, and communication scheduling must be performed at run-time.

4.1 Communication Scheduling

The PRL provides a set of generic communication scheduling routines through function templates over *Distribution* classes. We define communication scheduling to be a transformation from a communication pattern to a communication schedule, with respect to the corresponding *Distribution* classes. Communication patterns describe the required data motion by global subscripts, while the communication schedules describe the required data motion by local subscripts. Expression of communication patterns is simple, because it does not involve data distribution details. Therefore, we require users to express communication patterns in a problem-oriented way, and the PRL provides run-time support to generate the corresponding communication schedules. In this way, communication scheduling provides the encapsulation of data distribution details related to communication.

Different communication scheduling routines are provided according to different communication patterns: one-to-one, one-to-many(e.g. gather or reduction), and many-to-one(e. g. scatter or expansion). For each pattern, there are three possible representations: enumerated, functional, and dimensional.

```
// Communication scheduling
// cp is a one-to-many communication pattern in a functional
//     representation
// cs is the communication schedule for the communication from
//     y to y defined by cp
   Sparse_Array  x;
   Mapping       CY_BLOCK;
   Distribution  y(x, CY_BLOCK);
   Comm_Pattern  cp = {<i,j> => <i,j+1>,<i,j-1>,<i-1,j>,<i+1,j>};
   Comm_Schedule cs = communication_scheduling(y, y, cp);
```

The communication scheduling can be implemented in parallel by using the run-time support provided by data distribution descriptors. There are two kinds of implementation schemes: sender- or receiver-initiated. At the local domain, sender/receiver-initiated communication scheduling collects all local subscripts to be sent/received to/from other domains, and all global subscripts to be received/sent from/to other domains. It then transfers the collected global subscripts to the corresponding receivers/senders, and finally transforms the received global subscripts to local subscripts.

The above generic implementation is scalable, if a good mapping strategy exploits locality. More than that, the results of communication scheduling can be reused (e.g. within a loop), because the communication scheduling is only dependent on data distribution descriptors, not on distributed objects. A lot of scientific and engineering problems are solved by so-called iterative methods in which communication schedules can be reused naturally.

4.2 Data Motion

In the PRL, three kinds of data motion routines that are generic with respect to *Disobj* classes and a communication schedule are provided, for one-to-one, one-to-many, and many-to-one communication patterns (schedules). In these routines, communication with computation can be overlapped by providing a user-definable function object. The PRL supports the overlapping of communication and computation by the owner-computation rule. A user-definable function objects defines operations between the target data elements and received source data elements after communication in a flexible way. For example, we can define = or += for one-to-many communication, and +=, *=, max or min for many-to-one communication.

```
// Overlapped communication and computation by reuse of
// communication scheduling
   Disobj<Distribution, double> tar(y), src(y);
   ADD_OP fun;                // function object for reduction
   for(int i=1; i<No_Iter; i++)
        data_motion(tar, src, cs, fun);
```

The data motion routines are provided as function templates which are generic with respect to communication schedules. Special or optimized communication scheduling can be implemented and then their results can be used by these data motion routines, if the resulting communication schedules are conform with the uniform type interface defined by the PRL.

The data motion routines encapsulate message passing details such as creation of a communication buffer and its management, communication and synchronization, and data packing and unpacking. They are implemented by an underlying message-passing library. It provides asynchronous send and receive operations, synchronization operations, and buffer management. Currently, the PRL is implemented on top of MPI, PVM and some native communication packages respectively.

5 Performance

Our library is developed in the PROMOTER[2] project and is used as a user-level library, and a run-time library for the PROMOTER compiler. The system was developed on our testbed system MANNA, and has been ported to IBM SP/2, CRAY T3E, HITACHI SR2201, and to a cluster of workstations using either Ethernet or the Myrinet communication hardware.

Within the project, many applications (finite element methods, heat conduction, computational fluid dynamics, elasticity) have been tested. We have also implemented some benchmark programs using the run-time library, and have compared the performance of our implementation with hand-written and optimized implementations based on MPI.

Table 1 shows the time of the CG and FFT programs in the NASPAR benchmarks running on the HITACHI SR2201 machine. The CG program has sparse matrices, the FFT program has only dense arrays.

Table 1. Wall-clock run time of NASPAR benchmarks (seconds)

Benchmarks	8 PEs	16 PEs	32 PEs	64 PEs
CG (A/B)	28.04/ 814.54	16.83/437.03	9.90/223.96	
FFT (A/B)		35.88/79.50	18.02/39.23	9.399/20.69

The performance is achieved by applying the specialization and reuse of communication schedules on the programs using the PRL. Our approach achieves approximately 90%-100% of the performance of the hand-written benchmarks that directly use MPI.

6   Related Work

Much work has been carried out in run-time support for data-parallel applications. One of the pioneering efforts in run-time support for data-parallel applications is the development of a series of run-time libraries: MULTIBLOCK PARTI, CHAOS, and CHAOS++ [1]. They have three different interfaces for dense arrays and sparse arrays with regular and irregular data distributions. They use virtual functions in the implementation. The PRL has a uniform interface and uses only class templates and function templates in a way that their invocations can be resolved at compile-time.

There are many parallel C++ efforts such as ICC++ [3], C** [4], PC++ [5], MPC++[6] and, HPC++[7], which must also deal with run-time support for data-parallel applications. Usually, only regular dense arrays are considered. In the ILLINOIS CONCERT SYSTEM [3], ICC++ expresses (irregular) data parallelism as task-level concurrency. The PRL not only expresses regular and irregular data parallelism, but also expresses task-level concurrency as data parallelism. For example, a tree or a graph can be expressed by a sparse array by assigning a unique subscript to each node. This results in easy generation of collective communication, which can be achieved in the Illinois Concert System only by comprehensive compile-time analysis.

Another similar work is POOMA [8]. The POOMA framework is constructed in a layered fashion, in order to exploit the efficient implementation on the lower levels, while preserving an interface germane to the application problem domains at the highest level. However, POOMA is not a general-purpose programming environment, because it is motivated by specific applications. The PRL is designed to be a generic-purpose user-level library for data-parallel applications, and also as a run-time library to support the PROMOTER compilation system for a general-purpose data-parallel language.

7 Conclusions

The PRL provides object-oriented run-time support for regular dense and irregular sparse structures for an easy and efficient data-parallel programming with the following features. Data distribution details are decoupled from data by introducing data distribution descriptors. Data distribution descriptors themselves are distributed for scalability. They have a uniform interface for both, dense and sparse arrays, allowing a generic implementation of distributed data.

With the help of these descriptors, operations on distributed data can be easily mapped from a specification through global subscripts (sequential execution) to a specification through local subscripts (SPMD execution) by work distribution and communication scheduling.

Descriptors, work distribution and communication scheduling are provided in generic form for fast prototyping, and can be optimized or specialized for an efficient implementation. Descriptors can be shared, and results of work distribution and communication scheduling can be reused at run time for efficiency.

References

[1] C. Chang, J. Saltz, and A. Sussman. CHAOS++: A runtime library for supporting distributed dynamic data structure. Technical Report CRPC-TR95624, Center for Res. on Parallel Computation, Rice University, nov 1995.

[2] W. K. Giloi, M. Kessler, and A. Schramm. PROMOTER: A high level object-parallel programming language. In *Proc. Int.Conf. on High Performance Computing*, New Delhi, India, dec 1995.

[3] A. A. Chien and J. Dolby. The Illinois Concert system: A problem-solving environment for irregular applications. In *Proc. DAGS'94, Symp. on Parallel Computation and Problem Solving Environments*, 1994.

[4] L. R. Larus. A large-grain, object-oriented data-parallel programming language. In U. Banerjee, A. Nicolau D. Gelernter, and D. Padua, editors, *Languages and Compilers for Parallel Computing (5th International Workshop)*, pages 326–341. Springer-Verlag, aug 1993.

[5] A. Malony, B. Mohr, D. Beckman, D. Gannon, S. Yang, F. Bodin, and S. Kesavan. A parallel C++ runtime system for scalable parallel systems. In *Proc. Supercomputing'93*, pages 140–152. IEEE CS Press, nov 1993.

[6] S. Matsuoka, A. Nikami, H. Ogawa, and Y. Ishikawa. Towards a parallel C++ programming language based on commodity object-oriented technologies. In *Proc. ISCOPE'97*, pages 81–88. Springer-Verlag, dec 1997.

[7] G. HPC. HPC++ white paper. Technical Report CRPC-TR95633, Ctr. for Res. on Parallel Computation, Rice University, 1995.

[8] J.V.W. Reynders et al. POOMA: A framework for scientific simulation on parallel architectures. In Gregory V. Wilson and Paul Lu, editors, *Parallel Programming using C++*, pages 553–594. MIT Press, 1996.

Component Architecture of the Tecolote Framework*

Mark Zander, John Hall, Jim Painter, Sean O'Rourke

Los Alamos National Laboratory, Los Alamos, NM, USA

Abstract. Los Alamos National Laboratory's Tecolote Framework is used in conjunction with other libraries by several physical simulations. This paper briefly describes the design and use of Tecolote's component architecture. A component is a C++ class that meets several requirements imposed by the framework to increase its reusability, configurability, and ease of replacement. We discuss both the motives for imposing these requirements upon components and the means by which a generic C++ class may be integrated into Tecolote by satisfying these requirements. We also describe the means by which these components may be combined into a physics application.

1 Introduction

Los Alamos National Laboratory's Blanca project is part of the Department of Energy's Accelerated Strategic Computing Initiative (ASCI), which focuses on science-based nuclear weapons stockpile stewardship through the large-scale simulation of multi-physics, multi-dimensional, stockpile-relevant problems. Blanca is the only Los Alamos ASCI project written entirely in C++. Tecolote, the underlying framework for the development of Blanca physics codes, provides an infrastructure for combining individual component modules to create large-scale applications that encompass a wide variety of physics models, numerical solution options, and underlying data storage schemes, activating only essential components at run-time [1]. Tecolote maximizes code re-use and separates physics from computer science as much as possible. This allows physics model developers to use the Parallel Object-Oriented Methods and Applications (POOMA) framework, upon which Tecolote is layered, to write algorithms in a style similar to the problem's underlying computational physics equations [2].

POOMA contains architecture and parallelism abstractions that allow the user to write parallel physics codes without worrying about the underlying architecture or communications libraries. POOMA provides C++ fields that are similar to Fortran-90 arrays, but have additional features, including domain decomposition, load balancing, communications, and compact data storage. POOMA's unique capabilities provide the methods developer with powerful tools for expressing various mesh types and multiple dimensions; this allows application

---

* This work was performed under the auspices of the U.S. Department of Energy by Los Alamos National Laboratory under Contract No. W-7405-Eng-36.

developers to write mesh- and dimension-independent physics code whenever possible. Combined with Tecolote, POOMA's flexibility allows us to keep pace with the ever-changing ASCI environment, rapidly prototype ideas, and build on what others have done rather than using valuable time to reimplement the same basic models on different architectures.

Tecolote is portable to all ASCI-relevant hardware, making full use of its available parallelism. By supporting the rapid implementation of physics models and their immediate application to problems on the ASCI scale, Tecolote provides a powerful and flexible run-time environment that allows users to create and compose physics codes with varying capabilities "on the fly."

2 Approach

As we will discuss later, the Tecolote Framework supplies an application programmer interface that supports factorization of applications into components. Factorizing an application enhances the programmer's ability to cleanly separate interfaces from implementations, encapsulate conceptually independent subparts of a program, avoid code duplication, and maximize code reuse. After factorization, the user can integrate desired components using techniques supplied by Tecolote. Key concepts supported by this component architecture include separation of computer science from physics in simulations, implementation-independence of component interfaces, and increased run-time configurability (through an input-file scripting language). This flexible approach is made possible through the use of C++ inheritance and virtual function polymorphism. However, a Tecolote component's increased modularity comes at a price: creation of and communication between components can be more expensive than corresponding operations for ordinary C++ objects. Thus, the components' granularity must be large enough that these operations do not impact the code's efficiency.

Tecolote facilitates factorization through several mechanisms:

- uniform run-time data-sharing interface (the `DataDirectory`) with dynamic scoping rules;
- facility for the run-time description of components' type and inheritance relations (`MetaType`) and the registration of this information in a single type table (`MetaSet`);
- means of configuring both data values and control flow without recompilation (the input file scripting language); and
- separation of I/O and computation through the designation of "persistent" data intended for I/O.

The remainder of this paper illustrates the above-described mechanisms and their interaction through the extended example of a gamma-law equation of state (EOS) model.

3   Setting Component Parameters - Persistents

In object-oriented programming, I/O methods are generally encapsulated in application classes. To adopt a new I/O format, whether binary instead of ASCII text or eight-digit instead of six-digit floating-point output, every application class must be modified.

In Tecolote, we separate what is needed for I/O from how I/O is executed. The "what" is specified by the application programmer in a persistent list that contains the class data members available for I/O. Another component, an I/O module, determines how I/O is actually executed. The I/O module knows how to extract persistent locations from objects' MetaTypes and how to perform some type of I/O operation. There may be several different I/O modules in a system, each corresponding to a different data format.

For example, GammaLaw class members are

```
REAL                    pmin;       // minimum pressure
REAL                    gamma;      // adiabatic gamma
```

The persistents are listed outside the class declaration:

```
template< class C >
BEGIN_PERSISTENT( GammaLaw< C > )
    PERSISTENT( REAL, pmin, "pmin" )
    PERSISTENT( REAL, gamma, "gamma" )
END_PERSISTENT
```

Persistents support factorization by separating I/O modules from application modules and by deferring decisions about data initialization until run-time. Factoring I/O out of application objects ensures consistent input and output formatting with little burden on the application programmer; it also localizes changes required to support new data formats.

4   Sharing Fields Between Components - DataDirectory

Two models may use different (or the same) fields to compute their respective results. However, because we want to use virtual-function polymorphism to call the two models interchangeably, the models must use the same calling sequence in their respective evaluation functions. To avoid passing different fields in the argument lists of the evaluation functions, we have developed another alternative: passing a single data structure to the EOS model constructor, which then holds all the fields needed for a material. This data structure is termed the DataDirectory.

A DataDirectory is actually just like any other Tecolote component, except that it may have any number of persistents with any names. In contrast, ordinary components may only contain the persistents specified in their persistent lists. Any object, including another DataDirectory, may be placed inside a `DataDirectory`. Thus the DataDirectory structure is hierarchical, much like a Unix directory structure. Two entries are automatically put in a newly created `DataDirectory` to allow traversal of the hierarchy: "Root," which points to the `DataDirectory` at the top of the hierarchy, and "Parent," which points to the immediate predecessor in the hierarchy of the current `DataDirectory`.

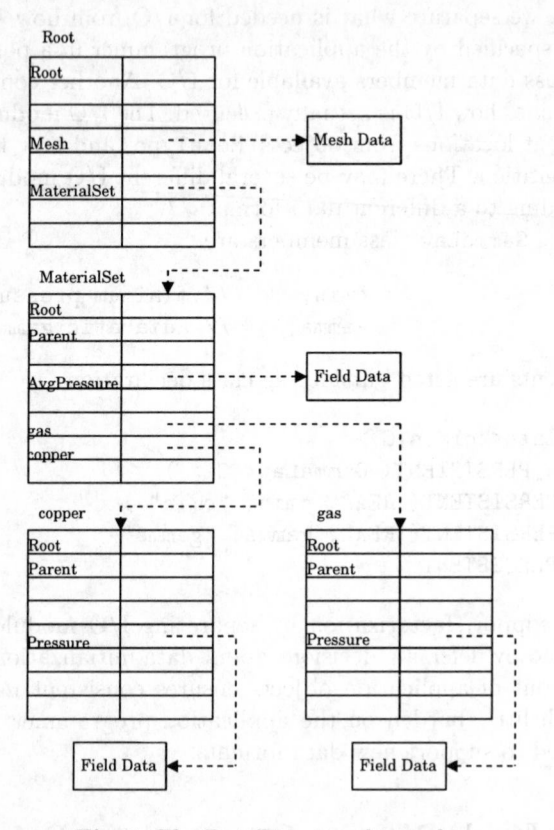

Fig. 1. The DataDirectory hierarchy

Figure 1 shows the DataDirectory hierarchy used for a multi-material simulation (for simplicity, only a few POOMA Fields are shown here). Unlike a Unix file structure, the DataDirectory structure has scoping rules similar to those of C++ inheritance. However, whereas C++ scope is determined by compile-time class inheritance relations, DataDirectory scope is determined by run-time object nesting. For instance, when `GammaLaw` attempts to get the `PhysicsMesh` from the "Material" `DataDirectory`, it fails to find it. Therefore, the search continues up the hierarchy, examining the "Material Set" and "Root" directories and terminating after it finds the requested `PhysicsMesh` in the "Root" directory.

The example below, from the `GammaLaw` class, illustrates the use of the `DataDirectory` macro GET:

```
ScalarField<C>& IntEnergy(
        GET("IntEnergy", Mat, ScalarField<C>, (Mesh))
);
```

The first argument is the name of the requested item; the second is the `DataDirectory` in which the search starts (`Mat`[erial] is a `DataDirectory`); the third is the type of the `DataDirectory` item; and the fourth (if present) represents the constructor arguments that are needed if the item is not present and must be added to the `DataDirectory`. The `GET` macro returns a reference to the object found in the `DataDirectory`.

The `DataDirectory` improves code factoring because a single data structure is passed to methods that otherwise would use different calling sequences. By deferring the association of data with a particular model until its actual instantiation, the DataDirectory structure provides generalized parameter-passing without explicit parameter lists. It enhances integration by supplying a mechanism for transparent data sharing among independent modules.

5 Building Components from an Input File - Scripting

At the start of its execution, a Tecolote program must specify which modules to use, their initial data values, and the high-level control structure in which they are applied. Tecolote uses a different component for each option and employs persistents to fill in the data needed by the options. Therefore, a Tecolote program built from components must use a methodology that creates objects from its components and places persistent data in the objects. We incorporate a scripting language into the Tecolote framework to accomplish these tasks.

Each object is described by an object name, a `MetaType` name, and a list of persistent values. Object hierarchy (or nesting) is indicated by listing one object in the persistent list of another. The following example shows nested objects where a `GammaLaw` is created as the `Eos` persistent of a `Material`. An object's constructor is called before its persistents are loaded. Therefore, an optional initialize function can be called after an object's persistents have been loaded to perform further initialization that is dependent on persistent values.

```
gas = Material(
    Eos = GammaLaw(
        gamma = 0.5,
        pmin = 0.001
    )
),
```

In both debugging and actual use, it is desirable to change the control flow of a program without rebuilding the entire code. Tecolote provides this flexibility by allowing the user to specify higher level function sequences in the input file. The necessary facilities are already provided by Tecolote and may be extended by an application programmer through a `MetaType` that maps methods and functions into function objects available from the input file.

The language, based on Backus' FP language [3] and Robison's IFP [4], includes program-forming operations (PFOs), basic objects, and elementary operations. The PFOs include control structures such as branching (If), looping (While), and sequences (Compose). Examples of basic objects are Lists, Numbers, and strings. Elementary operations that act on basic objects might include arithmetic, comparison, and collection functions such as concatenation and length. In Tecolote, as in FP, users may define new objects and functions, but not new PFOs.

The input file describes the initial object hierarchy of a program. In addition, it selects which components will be used as well as the initial values of their persistents. The functional scripting language allows the programmer, by combining function objects using PFOs, to define an application's high-level behavior.

6   Registering a Component with Tecolote - MetaTypes

In most applications, a module must directly reference other modules with which it interacts. This requirement obstructs factorization and prevents the application programmer from deferring module interactions until run-time. In addition, it is tedious to find all references to a module when replacing a module referenced in many locations within the program. In contrast, Tecolote components are registered only once – in the MetaSet, a table containing all the program components (see Fig. 2). Individual components interact only indirectly, through the MetaSet, promoting component independence.

A module is registered as a component with Tecolote by using a MetaType which, when invoked, automatically registers itself with the MetaSet. An object's MetaType is an object in its own right, much like Java's "Class" class [5]. The MetaType for a class:

- has a name in the MetaSet,
- holds the persistent list for that class,
- can create and initialize that class,
- carries its C++ type information to run-time, and
- can convert the MetaType from or to a single base class.

Many languages create a unified type hierarchy by having a single base type from which all other classes are derived (Java's Object class, for example). Tecolote classes, on the other hand, do not share a common base class. Eliminating this restriction allows Tecolote to incorporate classes not written for the framework (such as standard container classes and C++ basic types). Non-Tecolote classes are incorporated into the Tecolote type system by describing their basic features and their persistents in a MetaType.

In the example below, the GammaLaw<Cell> class is registered with the framework.

```
#include "GammaLaw.hh"
static MetaTecolote<GammaLaw<Cell>, Eos>
    GammaLawMeta("GammaLaw", MAKE_PERSISTENTS(GammaLaw<Cell>));
```

The generic `MetaTecolote` class is used to instantiate objects of classes with a constructor that takes `DataDirectory` and `string` arguments. Other `MetaTypes` are available that instantiate objects of classes with a constructor that takes no arguments. `MetaTecolote` is best used for classes that need more information about their environment, while other `MetaTypes` are used for classes written without knowledge of Tecolote. The class `GammaLaw<Cell>` is given the name "GammaLaw" in the `MetaSet` and has the base class `Eos`. The `MAKE_PERSISTENTS` macro registers the persistent information defined for the GammaLaw class.

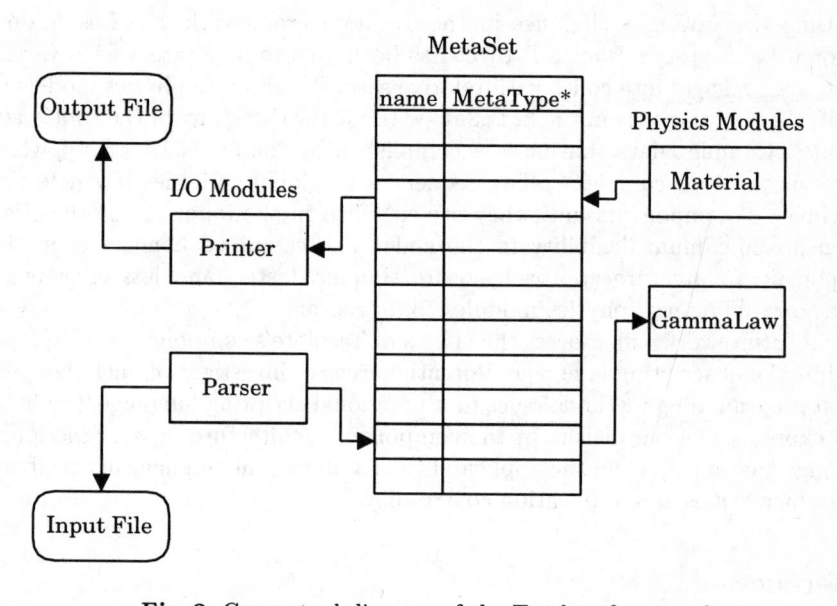

Fig. 2. Conceptual diagram of the Tecolote framework

All `MetaTypes` in the system register themselves with the `MetaSet`, a table that may be searched either by the `MetaType` name or by the type of the class that the `MetaType` contains. Although all modules are registered identically, physics modules and I/O modules interact with the `MetaSet` differently. Physics modules do not make explicit use of the `MetaSet`; however, I/O modules use the `MetaSet` explicitly both to build object hierarchies from input and to output program data. One example of this interaction, mentioned above, is the use of persistent information to perform output. Input modules also use a `MetaType`'s constructor, initializer, and base class in creating the object hierarchy.

Two I/O modules shown in Fig. 2 are the `Parser` and the `Printer`. The `Parser` reads data from the input file and then finds the `MetaType` by name in the `MetaSet`. After finding the `MetaType`, the `Parser` creates the object, loads its persistents, and calls its initializer. When an object is passed to the `Printer`, the `Printer` uses its C++ type to find the corresponding `MetaType` in the `MetaSet`.

It then prints the `MetaType` name, scans the persistent list, and prints each sub-object in turn.

`MetaTypes` and the `MetaSet` are key elements for supporting factorization in Tecolote. `MetaTypes` turn C++ classes into components, and the `MetaSet` may be searched for any component in an application program, which provides a level of indirection between components.

7 Conclusion

By using the various techniques in the Tecolote Framework, the Los Alamos National Laboratory's Blanca Project has been able to integrate a wide variety of physics packages into codes with relative ease. We add new physics models by modifying only the code in the `MetaSet`, without rewriting any of the code's I/O modules. Complex data sharing is accomplished by the `DataDirectory`, which allows us to avoid complex calling sequences or global variables. By deferring the choice of components until they are specified in the input file at run-time, we ensure maximum flexibility in the code. The combined benefits from this component architecture approach ensure simpler, faster, and less error-prone means for adding new physics modules to a program.

Our future work will address the effects of Tecolote's component architecture and functional scripting lanugage. Potential areas of investigation include application programming methodologies in a functional scripting language, the high-level expression of parallelism in the component architecture (given functional-language guarantees), and the applicability of a functional language in specifying object hierarchies and application control flow.

References

[1] K.S. Holian, L.A. Ankeny, S.P. Clancy, J.H. Hall, J.C. Marshall, G.R. Mcnamara, J. W.Painter, and M.E. Zander. TECOLOTE, an Object-Oriented Framework for Hydrodynamics Physics. In *Proceedings of the Conference on Numerical Simulations and Physical Processes Related to Shock Waves in Condensed Media*, Oxford, England, September 1997.

[2] J. V. W. Reynders, J. C. Cummings, P. J. Hinker, M. Tholburn, S. Banerjee, M. Srikant, S. Karmesin, Atlas S., K. Keahy, and W. F. Humphrey. *POOMA: A Framework for Scientific Computing Applications on Parallel Architectures*. MIT Press, 1996.

[3] J. Backus. Can Programming Be Liberated from the von Neumann Style? A Functional Style and its Algebra of Programs. In *Communications of the ACM*, pages 613–641, August 1978.

[4] A. Robison. Illinois functional programming: A tutorial. *BYTE*, pages 115–125, February 1987.

[5] J. Gosling, B. Joy, and G. Steele. *The Java Language Specification*. Addison-Wesley, August 1996.

Parallel Object Oriented Monte Carlo Simulations

Matthias Troyer[1,2], Beat Ammon[2], and Elmar Heeb[2]

[1] University of Tokyo, Tokyo, Japan
[2] ETH Zürich, Zürich, Switzerland

Abstract. We discuss the parallelization and object-oriented implementation of Monte Carlo simulations for physical problems. We present a C++ Monte Carlo class library for the automatic parallelization of Monte Carlo simulations. Besides discussing the advantages of object-oriented design in the development of this library, we show examples how C++ template techniques have allowed very generic but still optimal algorithms to be implemented for wide classes of problems. These parallel and object-oriented codes have allowed us to perform the largest quantum Monte Carlo simulations ever done in condensed matter physics.

1 Introduction

The Monte Carlo method[1] has been one of the most successful, if not *the* most successful numerical method in simulation of physical systems. Its applications span all length scales, ranging from large astrophysics simulations of galaxy clusters, to simulation of properties of solids and liquids, down to simulations of quarks and gluons, the constituents of protons and neutrons.

In solid state physics usual Monte Carlo algorithms were easy to vectorize and ideally suited for vector supercomputers. However, in the most interesting cases, close to phase transitions, these "local" Monte Carlo algorithms suffer from so-called "critical slowing down," which leads to an extra factor of L^2 in the CPU-time (L is the system size). Modern "cluster" algorithms [2, 3] beat this slowing down, but one has to deal with much more complex data structures and with algorithms that do not vectorize well.

In this paper we present how almost all kinds of Monte Carlo simulations, including the cluster algorithms, can be parallelized very efficiently and introduce a Monte Carlo class library and application framework that automatically performs this parallelization. Additionally, we present our experiences in using C++ template techniques to write generic Monte Carlo programs for a wide class of model systems, and in using them for more than 600 years of CPU time on a wide variety of workstation clusters and massively parallel machines.

2 Monte Carlo Simulations

Monte Carlo simulations are the only useful way to evaluate high-dimensional integrals. Such integrals are very common in the simulation of many-body systems.

For example, in a classical molecular dynamics simulation of M particles, the phase space has dimension $6M$ (3 coordinates each for positions and velocity).

Usual numerical integration techniques are very slow for high-dimensional integrals. For example, with the Simpson rule in d dimensions with N equidistant points, the error decreases as $N^{-4/d}$. For the corresponding Monte Carlo summation with N random points x_i sampled with some distribution $p(x_i$ the integral is estimated by $\int f(x)dx = \frac{1}{N} \sum_{i=1}^{N} f(x_i)/p(x_i)$ and the statistical error decreases as $N^{-1/2}$. For $d > O(10)$ the Monte Carlo integration method is thus faster.

Usually the points x_i are a Markov process $x_1 \to x_2 \to \ldots \to x_i \to \ldots$. Starting from a random configuration x_1 the Markov process must be iterated for a certain number N_{eq} of equilibration steps before it produces N random samples having the correct probabilities. This will be important for the performance of the parallel implementation. For more details about Monte Carlo methods, especially "importance sampling" and other techniques for reducing the statistical error, we refer to standard textbooks[1].

3 Parallelization and Performance of Monte Carlo Simulations

Our typical Monte Carlo simulations are easy to parallelize at several levels of granularity:

- Often we need many simulations for hundreds of different parameter sets (system sizes, temperatures, and so forth). Being independent, they can be parallelized trivially, with negligible overhead and almost perfect scaling, as little inter-processor communication is needed. For example, we found a speedup of 95.5 on 96 nodes of an Intel Paragon. For numbers of simulations larger than the number of available nodes, this level of parallelization is efficient.
- For one Monte Carlo simulation, uncorrelated Markov chains $\{vecx_i\}$ of statistical samples can be generated on different nodes by starting independent Monte Carlo runs with different random seeds. This level of parallelization however incurs a slight overhead, since each run needs to be equilibrated individually. On P nodes this leads to a theoretical maximal speedup of $P(1 + N_{eq}/N)/(1 + PN_{eq}/N)$. Since typically $N \approx 100N_{eq}$ this level of parallelization scales well to 20 times more nodes than simulations.
- Only if the equilibration time N_{eq} is very long or if memory needs require it, is it worth parallelizing a single Monte Carlo run. For example, this is possible by distributing the particles in the simulation over different nodes. This is, however, rarely done because of communication overhead.

We found, however, that the main bottleneck in scaling to a large number of nodes is caused by disk I/O needed at the beginning and end of each job (a simulation typically takes several weeks and thus has to be split into many separate jobs, requiring us to temporarily store the configurations on disk). Due to limitations of the parallel file system of the Hitachi SR2201 we used, this time grows faster than the data size, as we increase the number of nodes (250 sec for 256 nodes, 780 sec for 512 nodes, and 2970 sec for 1024 nodes for a typical quantum Monte Carlo simulation). On the machine we used the CPU time for a large job is unfortunately limited to one hour per node, so a typical program scales well only up to 256 nodes, where the disk I/O overhead is of the order of 15%.

4 The Alea Application Framework and Class Library

Monte Carlo simulations typically need a large amount of CPU time but fortunately, parallelize well. With the application framework and class library we have developed, many scientists with no experience in parallel computing can make use of the power of massively parallel computers for Monte Carlo simulations. The Alea library (Latin for "dice"), written in C++, automatically parallelizes many types of Monte Carlo simulation at the two generic levels mentioned above.

4.1 Classes for Monte Carlo Simulations

From a user's point of view, the library consists of three main classes, from which those specific to the Monte Carlo simulation are derived.

- A `simulation` class handles the parallelization of the different runs and the merging of the results of these runs. The user only has to override a `work()` member function, specifying the amount of work which needs to be done on this simulation. This value is then used for load balancing and serves as a termination criterion once it is zero.
- A `run` class implements the actual Monte Carlo simulation. The following functions have to be implemented for this class:
  - a constructor to start a new run
  - functions to access data in a dump, as discussed in Sec. 4.2 below.
  - a criterion `is_thermalized()` tells if the run is in equilibrium.
  - a function `do_step()` performs one Monte Carlo step and measurement.
- A container class, `measurements`, collects all the Monte Carlo statistics.

This is all the information the library needs to know about a specific Monte Carlo simulation. The library takes care of parameter input and startup, hardware independent checkpointing (see Sec. 4.2), parallelization (see Sec. 4.3) and dynamic load balancing, and evaluation and output of results

4.2 Object Serialization

An object serialization scheme was introduced to enable reading/writing of objects from/to data files, and transmission of objects to remote nodes.

C++ does not contain built-in object serialization, unlike Java. The C++ "iostream" library is also not suitable, since it is designed for text output and does not ensure that an object can be recreated from the textual output.

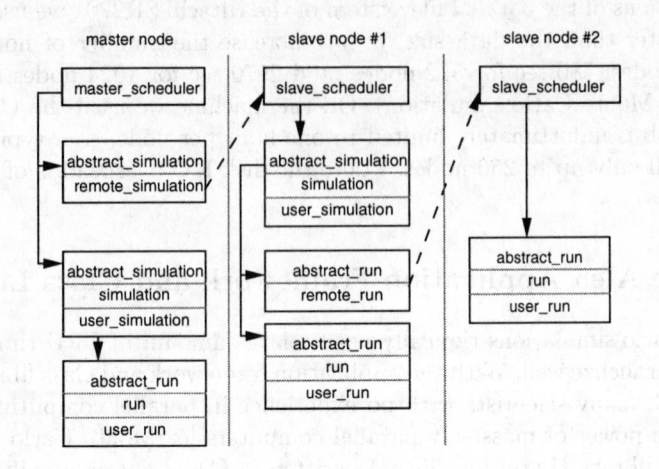

Fig. 1. Illustration of the parallelization and remote object creation. Each box represents an object. The labels of each object, from top to bottom, represent the class hierarchy from base class to most derived class. Solid lines are the creation of local objects, which in case of proxies send a message to a slave node to request the creation of the actual object. Illustrated is the creation of two simulations, which subsequently create three runs.

However, our implementation of object serialization is modelled after the "iostream" library. Objects can be written to odump streams using `operator <<` and read from idump streams using `operator >>`. Extensions to new classes are done just as in the iostream library, by overloading these operators. In particular, we have implemented two important types of such "dumps":

- `xdr_odump` and `xdr_idump` use the XDR format to write the data in a hardware independent binary format. These are used for hardware independent checkpoint files and for storing results of simulations.
- `mp_odump` and `mp_idump` which use an underlying message passing library to send objects from one node to another.

These latter classes allow easy parallelization using distributed objects.

4.3 Parallelization Using Distributed Objects

Simulations are parallelized as discussed in Sec. 3. The master node determines how much work needs to be done by each simulation and distributes the simulations across the available nodes accordingly. It then creates the `simulation` objects remotely, which in turn create one or several `run` objects.

Remote object creation is done by creating a proxy object (called either `remote_simulation` or `remote_run`), which in turn sends a message to the remote node requesting the creation of the object. Remote method invocation similarly invokes the method of the proxy object, which then sends a message to the remote node requesting the invocation of the method, and perhaps waits for a return value. Figure 1 shows this class hierarchy and method invocation. The scheme is simplified greatly compared to general distributed object systems since on each node, there exists at most one `simulation` and one `run` object.

Slave nodes check for messages and perform the requested method invocations. If no message needs to be processed, they call the `do_step` method of the local Monte Carlo `run` object to perform Monte Carlo steps.

4.4 Support Classes

In addition to the Monte Carlo simulation classes, the library also provides a variety of useful classes for parameter input, Monte Carlo measurements and their error analysis, the analysis of time series and the generation of plots.

4.5 Failure Tolerance

The library was designed to be tolerant to failure of single workstations on a workstation cluster when using PVM as the underlying message passing library. This is important as it allows us to perform the calculations on workstations in environments where other users reboot machines. Failure recovery is implemented by period checkpointing and by automatically restarting failed simulations from the latest checkpoint. Since the C++ exception mechanism is not yet fully supported by compilers, the implementation of this feature of the library has been delayed until a future release.

5 Object-Oriented Techniques for Monte Carlo Simulations

The first version of the above Monte Carlo library was developed in 1994 and 1995. At that time, the performance of C++ for scientific simulations was not good enough to allow the use of object-oriented techniques for the CPU intensive parts of the actual Monte Carlo simulations. They were coded in C-style C++ or even in FORTRAN.

Meanwhile, the template mechanism has been extended and is supported by more compilers. The use of "light objects" [4] and expression templates [4, 5] allows a higher level of abstraction and the use of object-oriented design *without* any abstraction penalty in the performance.

In the past year we have, with good success, made extensive use of such template techniques to develop generic, but still optimal, algorithms for a variety of condensed matter problems, and have used these programs successfully for a large number of simulations. This is, to our knowledge, one of the first applications of such techniques to large high-performance numerical calculations.

5.1   Generic Simulations Using Templates

Simulations of condensed matter problems have to be done for a variety of crystal lattice structures. Thus, it is advantageous to write a simulation program for general lattice structures. Usually, this is done by storing the lattice as a two-dimensional array of all neighbors of all sites. Using modern C++ compilers, we can describe the lattice by a class, shown here for a chain of sites:

```
class chain_lattice {
public:
  typedef unsigned int site_number;
  chain_lattice(site_number l) : length(l) {}
  site_number volume() {return length;}
  site_number neighbors(site_number site) {return 2;}
  inline site_number neighbor(site_number site,site_number nb)
    {return (nb ? (site==0? length-1 : site -1)
                : (site==length-1 ? 0 : site+1));}
private:
  site_number length;
};
```

This class can then be used as a template parameter of the run class:

```
template<class LATTICE,class MODEL>class user_run: public run {
private:
  LATTICE lattice;
  MODEL model;
public:
  virtual void do_step(); ...
};
```

In the CPU-intensive part (the function do_step()), most of the time is spent evaluating an interaction energy or cost function like:

```
for (typename LATTICE::site_number i=0;i<lattice.volume();++i)
  for (typename LATTICE::site_number n = 0;
  n < lattice.neighbors(site); ++n)
  {
  ... model.interaction(state[i],state[lattice.neighbor(i,n)]) ...
  }
```

Implementing the lattice information through template parameters as inlined functions as above allows the compiler to optimize more aggressively. In this example the innermost loop can be unrolled, and no memory access is needed to determine the neighbor, in contrast to the typical FORTRAN implementation which stores the numbers of the neighbors in two-dimensional arrays.

Similarly, Monte Carlo algorithms for a wide class of models and systems often differ just by a function describing the interaction, or by a type representing the states. These functions are typically very simple, containing just a few

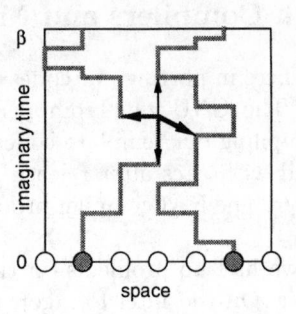

Fig. 2. Example of a world line configuration in a quantum mechanical simulation

operations. Obviously, a generic program using virtual function calls to evaluate an interaction energy is prohibitive due to an immense performance penalty associated with the virtual function call. On the other hand, providing the model as a template parameter, again allows generic, but highly optimized implementations. Details and examples will be presented in forthcoming publications.

5.2 Light Objects

Another use of templates and "light objects", which is maybe even more important, is for data structures representing a physical state. In the quantum Monte Carlo world line algorithm [3], a state is represented by "world lines" of particles, as shown in Fig. 2. These world line configurations are described by the position (horizontal axis) and times (vertical axis) of kinks in the world lines. In the Monte Carlo procedure, these kinks are shifted around. For that, it is necessary to know for each kink, the neighbors in the time direction (thin arrows in the figure), and to the previous kink on neighboring sites (thick arrows).

For simulations at low temperatures, where there are many such kinks along vertical lines, it is advantageous to store and update links to all these neighbors in a $(z + 2)$-fold linked list, where z is the number of neighbors.

At high temperatures, however, it is faster to store only the links along one spatial site (thin lines) and to find the spatial neighbors (thick lines) by searching along the linked lists at those sites.

Providing the actual representation of these kinks as a template parameter allows us to have optimized codes for both high and low temperatures available in the same program, which in other languages would only be possible using a preprocessor. Thus, being able to optimize and fine-tune the data structures easily has allowed us to get C++ codes that run faster than FORTRAN programs for the same problem (33500 moves per second compared to 29000 in [6]) Note that this is not due to inherent speed of C++ versus C or FORTRAN. It is caused by the fact that in practical complex applications, as compared to benchmarks, well coded C++ allows easier optimization of data structures.

6 Experiences with Compilers and Machines

We summarize our experiences in porting our codes to several parallel machines and workstation clusters. The GNU g++ compiler (version 2.7.2) sometimes caused problems when compiling our template codes with optimization, but no problems were encountered with egcs after release 1.0.2. We had no problems with the KAI C++ compiler. The KAI compiler produced code which performed about 20% better.

On parallel machines we had no problems on the Cray T3E, due to availability of the KAI compiler. On the Intel Paragon we used the GNU compiler successfully. The Hitachi SR2201, the fastest machine in the world at its introduction, gave us problems because there was limited template support in its C++ compiler.

In Monte Carlo programs, message passing speeds are irrelevant since almost no communication is necessary. The performance bottleneck is the I/O to the parallel file systems, at the beginning of a job to load the last configuration, and at the end to store the new configurations. Allowing large jobs to run for a longer time would enable us to extend the scaling beyond 256 nodes.

7 Summary and Applications

The library and programs discussed have been used now with success for three years, and have enabled us to perform the largest Monte Carlo simulations ever done for quantum mechanical simulations in condensed matter physics, sometimes three orders of magnitude larger than previous simulations. This in turn has allowed us to answer long-standing interesting questions in this field[1].

The library will be publicly available when we finish rewriting it to use the new standard C++ library. Interested persons can contact the authors to obtain the old version. In the future, we plan to extend template techniques to develop generic programs and classes also for other algorithms, methods, and quantum operators in the field of quantum simulations in physics.

References

[1] K. Binder and D. W. Heermann, *Monte Carlo Simulation in Statistical Physics : An Introduction*, Springer Series in Solid-State Sciences, 80 (Springer-Verlag, 1989).
[2] R.H. Swendsen and J.-S. Wang, *Phys. Rev. Lett.* 58, 86 (1987); U. Wolff, *Phys. Rev. Lett.* 62, 361 (1989).
[3] H.G. Evertz, in *Numerical Methods for Lattice Quantum Many-Body Problems*, ed. D.J. Scalapino, (Addison Wesley Longman, Frontiers in Physics, 1998).
[4] S.W. Haney, *Comp. in Phys.* 10, 552(1996); A.D. Robison, *ibid.* 10, 458 (1996).
[5] T. Veldhuizen, *C++ Report* 7, 26(1995); *ibid.* 7, 36(1995).
[6] K. Harada and N. Kawashima, Report cond-mat/9803090.

---

[1] For a selection of papers using the library see the papers coauthored by Frischmuth or Troyer from 1995-1998 in the cond-mat archive at http://xxx.lanl.gov/.

A Parallel, Object-Oriented Implementation of the Dynamic Recursion Method

Wolfram T. Arnold and Roger Haydock

University of Oregon, Eugene, OR, USA

Abstract. We present the first parallel, object-oriented C++ implementation of the dynamic recursion method. The recursion method is a means to tridiagonalize sparse matrices efficiently and is useful for a wide number of problems in physics. Dynamic recursion describes an optimization of the standard recursion method by operating only with a dynamically varying subset of basis vectors—reducing memory needs and allowing the computation of very large systems. We show how a graph-based data structure permits storing and multiplying sparse matrices and vectors efficiently. We use a tree structure to cope with the dynamically changing basis set, manifested by perpetual creation and elimination of vector components and matrix elements. A "workpile" approach is employed to allow thread-based parallel execution. Systems with up to 10^7 matrix elements have been simulated with the current implementation and the Anderson metal-insulator transition has been studied as a test-bed project.

1 Introduction

With the rising popularity of object-oriented languages for scientific applications, a number of excellent programming "environments," such as the Blitz++ [1] and POOMA [2] packages, have been developed which, as a result of their dedication to the object-oriented paradigm, offer data structures and operations resembling physical or mathematical concepts. This allows the straightforward implementation of scientific models while eliminating the need of machine-oriented thinking. POOMA, in addition, features completely encapsulated parallelism.

Following the same philosophy, we have developed a C++ kernel for applications that are based on the dynamic recursion method [3, 4]. Our parallel "recursion engine" is applicable to a variety of problems that can be formulated as sparse matrices such as those common in condensed matter physics. The application programmer only needs to specify a short module of code describing the physical system under study. The parallel recursion algorithm is captured in the data structures provided and completely transparent to the user.

2 Recursion Method

A large number of problems in physics can be formulated in terms of sparse matrices which have only a small number of non-zero elements. In fact, this is possible for any physical system with only short-ranged interactions in real space. Once the physical problem is formulated with sparse matrices, it is usually solved by a matrix transformation, which makes computational approaches attractive. While a diagonalization yielding eigenvalues and eigenvectors is useful for small systems, it is usually not the best choice for large systems, because diagonalizations are not only very resource-intensive, but the eigenvalues are also very sensitive to boundary conditions. Physically interesting quantities are distributions of eigenvalues, such as a projected density of states, which can be obtained from a tridiagonalization of the sparse matrix [3].

The recursion method [3] is an efficient technique for such a tridiagonalization. Suppose H is a large symmetric (Hermitian) sparse matrix. For a physical system, H represents the system's Hamiltonian in some suitably chosen basis set. Then, tridiagonalization amounts to the matrix transformation

$$U^{\dagger}HU = J, \quad \text{with} \quad J = \begin{pmatrix} a_0 & b_1 & & & 0 \\ b_1 & a_1 & b_2 & & \\ & b_2 & a_2 & & \\ & & & \ddots & b_n \\ 0 & & & b_n & a_n \end{pmatrix}, \tag{1}$$

where U is the unitary transformation matrix to be found.[1] The recursion method successively generates the column vectors $\{\mathbf{u}_n\}$ of the matrix U by a recurrence relation. Starting from some vector $\mathbf{u}_0$, $\mathbf{u}_1$ is computed by

$$H\mathbf{u}_0 = a_0\mathbf{u}_0 + b_1\mathbf{u}_1 , \tag{2}$$
$$\text{and for } n \geq 1 \quad H\mathbf{u}_n = a_n\mathbf{u}_n + b_{n+1}\mathbf{u}_{n+1} + b_n\mathbf{u}_{n-1} . \tag{3}$$

The coefficients a_n and b_n are determined by the requirement that the vectors $\{\mathbf{u}_n\}$ be normalized and orthogonal to each other:

$$a_n = \mathbf{u}_n^{\dagger}H\mathbf{u}_n , \tag{4}$$
$$b_{n+1}^2 = [(H - a_nI)\mathbf{u}_n - b_n\mathbf{u}_{n-1}]^{\dagger}[(H - a_nI)\mathbf{u}_n - b_n\mathbf{u}_{n-1}] , \tag{5}$$

where I is the identity matrix. This prescription specifies all vectors $\mathbf{u}_1 \ldots \mathbf{u}_{n+1}$, but not the start vector $\mathbf{u}_0$. The choice of $\mathbf{u}_0$ determines the state of projection for the projected density of states that can be computed from the a_n and b_n. For the mathematical details of the relation between the tridiagonal matrix J and the projected density of states, see [3].

The recursion method is computationally interesting for physical problems because the projected density of states converges exponentially fast with the number of recursions. In other words, the approximation of the (potentially infinite) system H, that is represented by J, is improving exponentially with the number of transformation vectors included. Even more generally, this convergence property makes the method rather insensitive to any errors, such as those

---

[1] We follow the convention of using the symbol $\dagger$ to denote the Hermitian conjugate. If all matrices are real, this is equivalent to the transpose, indicated by T.

arising from the finite precision of computer arithmetic, the loss of orthogonality of the basis vectors and others. Te [4] has shown that the total error will always remain bounded and not accumulate exponentially. The concept of *dynamic recursion* has grown from this idea: One can eliminate the smallest components in the vectors $\{\mathbf{u}_n\}$, thereby reducing memory use which, in turn, allows an increase in the maximally computable system size within the given resources, while the total error remains below a predetermined threshold. The effect is to "optimize" the basis set used in the transformation for the desired system size and available resources.

In this group, dynamic recursion has been applied to the investigation of a metal-insulator transition in an alkali adlayer on a surface [5] and the study of the Anderson transition in a three-dimensional system [6]. The latter has been a test-bed project for our new parallel implementation of dynamic recursion. We have simulated systems with 10^7 sites in initial runs—a number likely to increase once inefficiencies in the code have been rooted out.

3   Data Structure and Algorithm for Dynamic Recursion

An implementation of this dynamic recursion algorithm requires the following: (a) an efficient way to store the sparse matrix H and the associated vectors $\mathbf{u}_n$, allowing for dynamic generation and deletion of components; (b) an algorithm for the matrix-vector multiplication $H\mathbf{u}_n$ of Eq. (3), interfacing with this storage scheme; (c) a partitioning of the problem suitable for concurrent execution.

Let us consider a simple physical example: A 2-d square lattice, e.g. of atoms, with the sites ("nodes") representing atomic states and the connecting lines ("edges") the non-zero overlap integrals between nearest neighbors (Fig. 2, bottom) will result in a sparse matrix. Mathematically speaking, the nodes represent the set of basis vectors, in which the matrix is given, and the edges the matrix elements between them. If one chooses some scheme for enumerating the sites, e.g. spiraling outward from some center site, the system can be described as a matrix H with the row and column indices referring to site indices. The matrix will be sparse because per site (per row in the matrix) there are only very few neighbors (non-zero column entries in that row). For a physical system, the elements of H can usually be derived from some formula; for the purpose of this paper, we will take the diagonal elements as zero and the off-diagonal elements as constant. The goal of the recursion now is to generate new basis vectors $\mathbf{u}_n$. If the original basis vectors, for example, are given by the set

$$(1,0,0,0,\dots)^T,\ (0,1,0,0,\dots)^T,\ (0,0,1,0,\dots)^T,\ (0,0,0,1,\dots)^T,\ \dots \qquad (6)$$

then a new basis vector will have a number of components on the original basis vectors. If the zeroth vector of the new basis, $\mathbf{u}_0$, is chosen

$$\mathbf{u}_0 = (1,\,0,\,0,\,0,\,0,\,0,\,\dots)^T, \qquad (7)$$

then, from Eq. (2), $\quad \mathbf{u}_1 = \frac{1}{2}(0,\,1,\,1,\,1,\,1,\,0,\,\dots)^T. \qquad (8)$

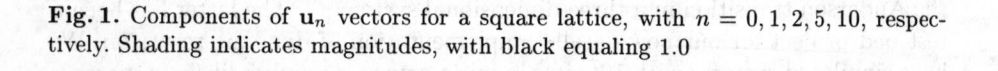

Fig. 1. Components of $\mathbf{u}_n$ vectors for a square lattice, with $n = 0, 1, 2, 5, 10$, respectively. Shading indicates magnitudes, with black equaling 1.0

Figure 1 illustrates this spatial propagation of the vector components for a 2-d square lattice. All sites that carry a non-zero component ("weight") are called "active". The ensemble of all active sites represents a given vector $\mathbf{u}_n$. Multiplication with H, i.e. $H\mathbf{u}_n$, will yield a vector in which all the neighbors to the active sites become active as well, and only the matrix elements between active sites are referenced. These matrix elements can be generated when needed and do not have to be stored at all times. In dynamic recursion, small elements of $\mathbf{u}_n$ can be deleted and so can their associated matrix elements.

A *graph* data structure mirroring the lattice, with nodes representing components in the vector $\mathbf{u}_n$ and edges the associated matrix elements, provides a natural way of storing the matrix and vector for the purposes of the $H\mathbf{u}_n$ operation which itself then amounts to a mere graph traversal (see pseudocode in Fig. 4). Along with the components of $\mathbf{u}_n$, the components of $\mathbf{u}_{n-1}$ are stored as well; they are needed for Eq. (3). We can see from Eqs. (4, 5) that per recursion step, two graph traversals are needed, because of the data dependency of Eq. (5) on a_n. In the pseudocode, these are referred to as first and second phase of the computation, respectively, with $H\mathbf{u}_n$ being calculated in the former.

The active sites make up a lattice that grows as a result of the matrix-vector multiplication. Wherever there are sites with fewer neighbors than the coordination number (four in the 2-d square lattice), new nodes are added. Some mechanism is then needed to determine if some of the existing nodes of the graph may qualify as neighbors to a new one (Fig. 2, bottom). The neighbor-finding problem can be solved by a tree structure associated with a binary subdivision of coordinate space (Fig. 2, top), as it has been used, for example, in the N-body problem [7]. At the same time, the tree contains the functionality of the graph data structure, with the graph traversal then amounting to a tree-traversal which can be done recursively in an efficient manner. In addition, a tree structure allows for straightforward parallel execution.

From an analytical viewpoint, the tree supplies an enumeration of the nodes. There is no unique way of doing this and the tree performs this enumeration in

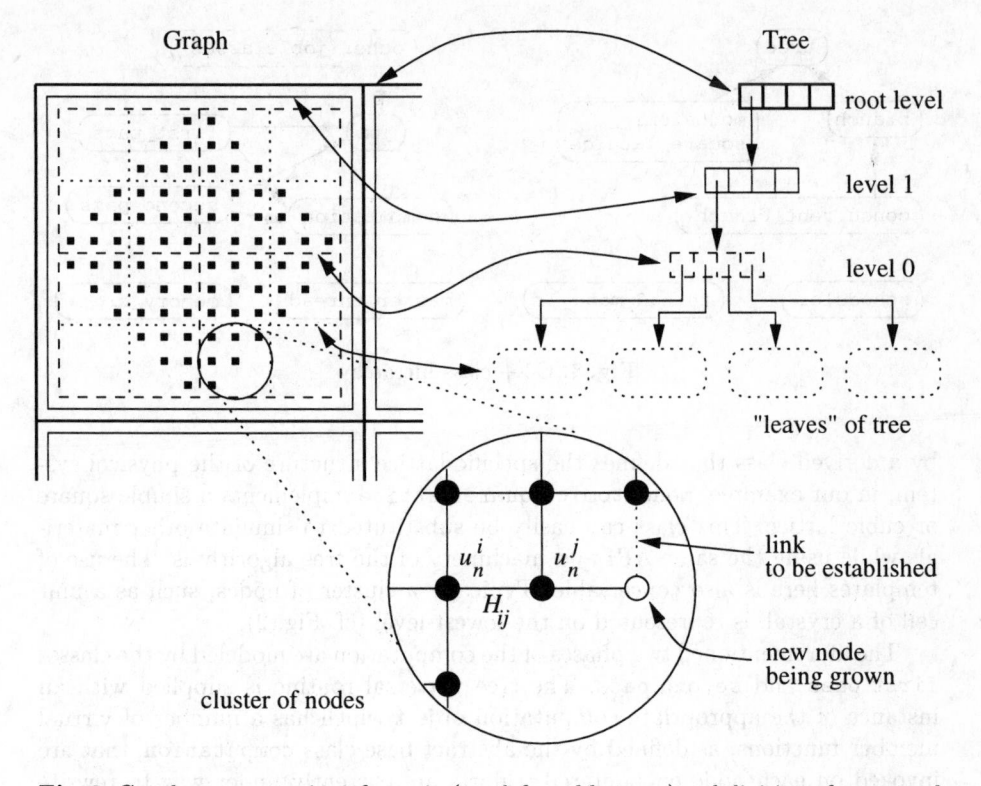

Fig. 2. Graph representation of matrix (top left and bottom); subdivision of space and correspondence to a tree structure (top left and top right); growth of a new node from an existing one and identification of neighbors (bottom)

such a way that the locality of the problem is preserved, i.e. nodes that are close together in space reside on adjacent branches of the tree. This makes neighbor-finding a local and thus fast process. At the same time, there is no performance penalty if the nodes are added and removed dynamically—a major advantage over other algorithms based on look-up lists or nearest neighbor maps.

4   C++ Class Design

The abstract concepts of graph, tree and nodes introduced in Sect. 3 lend themselves perfectly to an object-oriented implementation. C++ has been chosen as language because of its wide availability and high performance thanks to the latest generation of optimizing compilers such as KAI's C++ compiler [8].

Figure 3 shows the C++ design. The tree class hierarchy, headed by the abstract base class `tree`, is responsible for traversing the tree, finding neighbors and adding and deleting branches if this is called for as a result of nodes being created or killed. Intermediate levels in the tree are represented by `branch` objects, pointing to other `tree` objects below. The terminals of the tree are formed

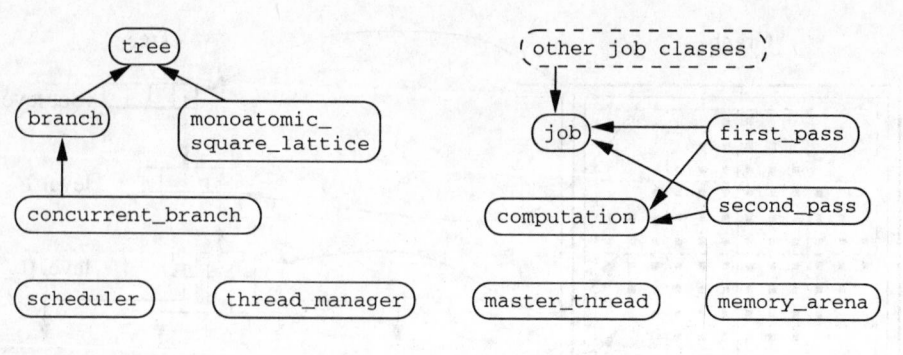

Fig. 3. C++ class hierarchy

by a derived class that defines the specific lattice structure of the physical system; in our example, monoatomic_square_lattice implements a simple square or cubic lattice. This class can easily be substituted to simulate other materials while using the same API and machinery of the tree algorithms. The use of templates here is also conceivable. Typically a cluster of nodes, such as a unit cell of a crystal, is represented on the lowest level (cf. Fig. 2).

The aforementioned two phases of the computation are modeled by the classes first_pass and second_pass. The tree traversal routine is supplied with an instance of the appropriate computation object which has a number of virtual member functions, as defined by the abstract base class computation, that are invoked on each node encountered. Efforts are currently under way to rewrite the code using templates to avoid the performance overhead of virtual functions.

The remaining classes in Fig. 3 perform certain auxiliary functions having to do with multi-threading and memory management. As dynamically allocated objects are of fixed size, heap management has been implemented very efficiently.

5 Parallelization Issues

For parallel execution, the computational problem needs to be divided up into concurrent sections. The tree traversal with neighbor look-up, representing the matrix-vector multiplication, is the computationally most expensive part in the recursion, but can easily be parallelized with different processors working on separate branches of the tree. Because the tree preserves locality, it offers inherent parallelism so that very little non-local data movement between processors is needed, minimizing communication bottlenecks.

These facts lend themselves to an efficient implementation based on threads in combination with a "workpile" paradigm [9]. We have used the POSIX pthread library [9, 10] which executes multiple threads concurrently on a multi-processor computer. In our code, the tree is partitioned into subtrees below a certain level, and pointers to these subtrees (instances of concurrent_branch) are held by instances of fist_pass or second_pass which are derived from the abstract base class job. All job objects are put on a workpile. A number of worker threads take

---

First pass of computation per recursion step to compute a_n, cf. Eq. (4)
   $a[n] \leftarrow 0$
   FOR $i = 0$ TO all nodes in the tree
      $sum \leftarrow u_n^i \times H_{ii}$
      FOR all neighbors u_n^j of u_n^i
         retrieve value u_n^j by neighbor-finding algorithm
         retrieve or generate matrix element H_{ij}
         $sum \leftarrow sum + u_n^j \times H_{ij}$
         IF neighbor u_n^j doesn't exist
            THEN grow new node, extend tree, add new branches as needed
      END FOR
      store sum for use in second pass
      $a[n] \leftarrow a[n] + u_n^i \times sum$, cf. Eq. (4)
   END FOR
Second pass of computation per recursion step to compute b_{n+1}, cf. Eq. (5)
   $b2 \leftarrow 0$
   FOR $i = 0$ TO all nodes in the tree
      $b2 \leftarrow b2 + [H\mathbf{u}_n]^i - a[n]u_n^i - b[n]u_{n-1}^i$, with $[H\mathbf{u}_n]^i$ stored from first pass
      update u_n^i, u_{n-1}^i, cf. Eq. (3)
   END FOR
   $b[n+1] \leftarrow \sqrt{b2}$, cf. Eq. (5)
Neighbor-finding algorithm, cf. Fig. 2
   $level \leftarrow -1$ (leaf level)
   WHILE u_n^i and u_n^j are in different branches
      go up in tree, $level \leftarrow level + 1$
      add coordinate of current octant to a traversal list
   END WHILE (common ancestor has been found)
   WHILE traversal list non-empty
      move down in tree, taking directions "conjugate" to the ones stored in list
      (e.g. if "south" was taken going up, "north" must be taken going down)
      remove item from traversal list
   END WHILE

Fig. 4. Pseudocode for three algorithms: first and second pass of computation, and neighbor-finding algorithm

jobs off the workpile and execute the associated operation, such as traversing the corresponding subtree. When the workpile runs empty, the coefficients a_n or b_n can be computed from the accumulated results of all threads. Other job classes exist that "prune" the tree by eliminating empty subtrees or perform other maintenance tasks. If the assignment of subtrees to processors were fixed, load imbalances would soon result from the dynamic growth of the tree. This difficulty arises in distributed computing models like MPI [11]. The latter are therefore harder to use efficiently for this problem. The thread-based approach, however, does require a shared memory architecture; the present program has been developed on a Silicon Graphics Power Challenge platform.

Another—rather subtle—problem arising in a shared memory multiprocessor architecture with cache is due to "false sharing." This is a cache line invalidation occurring when separate addresses contained in the same cache line are written to and read from by different processors. Although not a race condition, the penalty for that can be so severe that the program may run slower with multiple threads than with a single one. This problem, typical for multi-threading, is avoided by careful data alignment.

6 Conclusions and Outlook

We have presented a first parallel implementation of dynamic recursion in a C++ design which will allow ready adaptation to other problem domains, including interacting systems. Efforts are currently under way to define a clean API, possibly within a similar framework as the POOMA or Blitz++ systems.

The program in the present form has been used to study the Anderson disorder transition in a three-dimensional system [6]. Other work on metal-insulator transitions in two dimension is currently under way. The program, however, has a much wider range of applicability and it is hoped that in the future it may be used for other problems as well.

Acknowledgements Thanks are due to Sameer Shende, Janice Cuny, Duane Bailey, Allen Maloney, Chris Nex, Paul Bunson, John Conery and his computational science class '97. This work was supported by the National Science Foundation under grant nos. DMR-9319246, STI-9413532, CDA-9601802.

References

[1] Veldhuizen, T. et al. *Blitz++*, URL: http://monet.uwaterloo.ca/blitz/

[2] Reynders, J. et al. *Parallel Object-Oriented Methods and Applications (POOMA)*, URL: http://www.acl.lanl.gov/PoomaFramework/index.html

[3] Haydock, R. *The Recursive Solution of the Schrödinger Equation*, Solid State Physics, **35** (1980)

[4] Te, R.L., Haydock, R. Phys. Rev. B, **49**, no. 16, p. 10845 (1994)

[5] Haydock, R., Te, R.L. Phys. Rev. B, **57**, no. 1, p. 296 (1998)

[6] Arnold, W.T. *Investigation of the Localization of Electronic States in Three-Dimensional Systems with Anderson Disorder Using the Recursion Method*, Diploma Thesis, Universität Ulm, Germany (1998), URL: http://www.cs.uoregon.edu/~wolfram/

[7] Barnes, J., Hut, P. Nature **324**, p. 446 (1986)

[8] Kuck & Associates *KAI C++*, http://www.kai.com/C_plus_plus/_index.html

[9] Kleiman, S., Shah, D., Smaalders, B. *Programming with Threads*, SunSoft Press (1996)

[10] IEEE, *Information Technology–Portable Operating System Interface (POSIX)– Part 1: System Application: Program Interface (API), Threads Extension*, IEEE Standard 1003.1c-1995, also ISO/IEC 9945-1:1996

[11] *The Message Passing Interface*, URL: http://www.mcs.anl.gov/mpi/index.html

Object-Oriented Design for Sparse Direct Solvers*

Florin Dobrian[1], Gary Kumfert[1], and Alex Pothen[1,2]

[1] Old Dominion University, Norfolk, VA, USA
[2] NASA Langley Research Center, Hampton, VA, USA

Abstract. We discuss the object-oriented design of a software package for solving sparse, symmetric systems of equations (positive definite and indefinite) by direct methods. At the highest layers, we decouple data structure classes from algorithmic classes for flexibility. We describe the important structural and algorithmic classes in our design, and discuss the trade-offs we made for high performance. The kernels at the lower layers were optimized by hand. Our results show no performance loss from our object-oriented design, while providing flexibility, ease of use, and extensibility over solvers using procedural design.

1 Introduction

The problem of solving linear systems of equations $Ax = b$, where the coefficient matrix is sparse and symmetric, represents the core of many scientific, engineering and financial applications. In our research, we investigate algorithmic aspects of high performance direct solvers for sparse symmetric systems, focusing on parallel and out-of-core computations. Since we are interested in quickly prototyping our ideas and testing them, we decided to build a software package for such experimentation. High performance is a major design goal, in addition to requiring our software to be highly flexible and easy to use.

Sparse direct solvers use sophisticated data structures and algorithms; at the same time, most software packages using direct solutions for sparse systems were written in Fortran 77. These programs are difficult to understand and difficult to use, modify, and extend due to several reasons. First, the lack of abstract data types and encapsulation leads to global data structures scattered among software components, causing tight coupling and poor cohesion. Second, the lack of abstract data types and dynamic memory allocation leads to function calls with long argument lists, many arguments having no relevance in the context of the corresponding function calls. In addition, some memory may be wasted because all allocations are static.

We have implemented a sparse direct solver using different programming languages at different layers. We have reaped the benefits of object-oriented

* This work was partially supported by the National Science Foundation grants CCR-9412698 and DMS-9807172, by the Department of Energy grant DE-FG05-94ER25216, and by NASA under Contract NAS1-19480.

design (OOD) and the support that C++ provides for OOD, at the highest layer, and the speed of Fortran77 at the lower levels. The resulting code is more maintainable, usable, and extensible but suffers no performance penalty over a native Fortran77 code. To the best of our knowledge, this work represents the first object-oriented design of a sparse direct solver.

We chose C++ as a programming language since it has full support for object-oriented design, yet it does not enforce it. The flexibility of C++ allows a software designer to choose the appropriate tools for each particular software component. Another candidate could have been Fortran 90, but it does not have inheritance and polymorphism. We need inheritance in several cases outlined later. We also wish to derive new classes for a parallel version of our code. We do not want to replicate data and behavior that is common to some classes. As for polymorphism, there are several situations when we declare just the interfaces in a base class and we want to let derived classes implement a proper behavior.

In this paper we present the design of our sequential solver. Work on a parallel version using the message-passing model is in progress. Object-oriented packages for iterative methods are described in [1, 2].

2 Overview of the Problem

Graph theory provides useful tools for computing the solution of sparse systems. Corresponding to a symmetric matrix A is its undirected adjacency graph $G(A)$. Each vertex in the graph corresponds to a column (or row) in the matrix and each edge to a symmetric pair of off-diagonal nonzero entries.

The factorization of A can be modeled as the elimination of vertices in its adjacency graph. The factorization adds edges to $G(A)$, creating a new graph $G^+(A, P)$, where P is a permutation that describes the order in which the columns of A are eliminated. Edges in G^+ not present in G are called fill edges and they correspond to fill elements, nonzero entries in the filled matrix $L + D + L^T$ that are zero in A.

The computation of the solution begins thus by looking for an ordering that reduces the fill. Several heuristic algorithms (variants of minimum degree or nested dissection) may be used during this step. The result is a permutation P.

Next, an *elimination forest* $F(A, P)$, a spanning forest of $G^+(A, P)$, is computed. The elimination forest represents the dependencies in the computation, and is vital in organizing the factorization step. Even though it is a spanning forest of the filled graph, it can be computed directly from the graph of A and the permutation P, without computing the filled graph. In practice, a compressed version of the elimination forest is employed. Vertices that share a common adjacency set in the filled graph are grouped together to form supernodes. Vertices in a supernode appear contiguously in the elimination forest, and hence a supernodal version of the elimination forest can be used.

The factorization step is split in two phases: *symbolic* and *numerical*. The first computes the nonzero structure of the factors and the second computes the numerical values. The symbolic factorization can be computed efficiently using the supernodal elimination forest. The multifrontal method for numerical factorization processes the elimination forest in postorder. Corresponding to each supernode are two dense matrices: a *frontal matrix* and an *update matrix*. Entries in the original matrix and updates from the children of a supernode are assembled into the frontal matrix of a supernode, and then partial dense factorization is performed on the frontal matrix to compute factor entries. The factored columns are written to the factor matrix, and the remaining columns constitute the update matrix that carries updates higher in the elimination forest.

Finally, the solution is computed by a sequence of triangular and diagonal solves. Additional solve steps with the computed factors (iterative refinement) may be used to reduce the error if it is large.

When the coefficient matrix is positive definite, there is no need to pivot during the factorization. For indefinite matrices, pivoting is required for stability. Hence the permutation computed by the ordering step is modified during the factorization.

Additional details about the graph model may be found in [3]; about the multifrontal method in [4]; and about indefinite factorizations in [5].

3 Design of the Higher Layers

At the higher layers of our software, the goal was to make the code easy to understand, use, modify and extend. Different users have different needs: Some wish to minimize the intellectual effort required to understand the package, others wish to have more control. Accordingly, there must be different amounts of information a user has to deal with, and different levels of functionality a user is exposed to.

At the highest level, a user is aware of only three entities: the coefficient matrix A, the right hand side vector b, and the unknown vector x. Thus a user could call a solver as follows:

$$x = Compute(A, b),$$

expecting the solver to make the right choices. Of course it is difficult to achieve optimal results with such limited control, so a more experienced user would prefer to see more functionality. Such a user knows that the computation of the solution involves three main steps: (1) *ordering*, to preserve sparsity and thus to reduce work and storage requirements, (2) *factorization*, to decompose the reordered coefficient matrix into a product of factors from which the solution can be computed easily, and (3) *solve*, to compute the solution from the factors. This user would then like to perform something like this:

$$P = Order(A),$$
$$(L, D, P) = Factor(A, P),$$
$$x = Solve(L, D, P, b).$$

Here, P is a permutation matrix that trades sparsity for stability, L is a unit lower triangular or block unit lower triangular matrix, and D is a diagonal or block diagonal matrix.

At this level the user has enough control to experiment with different algorithms for each one of these steps. The user could choose a minimum degree or a nested dissection ordering, a left-looking or a multifrontal factorization. In addition, the user may choose to run some of the steps more than once to solve many related systems of equations, or for iterative refinement to reduce the error.

We organized the higher layers of our software as a collection of classes that belong to one inheritance tree. At the root of the tree we put the *Object* class, which handles errors and provides a debugging interface. Then, since the two basic software components are data structures and algorithms, and since decoupling them achieves flexibility, we derived a *DataStructure* class and an *Algorithm* class from *Object*. The first one handles general information about all structural objects and the second one deals with the execution of all algorithmic objects.

An important observation is necessary here. While full decoupling needs perfect encapsulation, the overhead introduced by some interfaces may be too high. Thus performance reasons forced us to weaken the encapsulation allowing more knowledge about several objects. For sparse matrices, for example, we store the data (indices and values) column-wise, in a set of arrays. We allow other objects to retrieve these arrays, making them aware of the internal representation of a sparse matrix. We protect the data from being corrupted by providing non-const access only to functions that need to change the data. Such a design implementation may be unacceptable for an object-oriented purist. However, a little discipline from the user in accessing such objects is not a high price for a significant gain in performance.

A user who does not want to go beyond the high level of functionality of the main steps required to compute the solution sees the following structural classes: *SparseSymmMatrix*, *Vector*, *Permutation* and *SparseLwTrMatrix*. The first class describes coefficient matrices, the second right hand side and solution vectors, the third permutations, and the fourth both triangular and diagonal factors. We decided to couple these last two because they are always accessed together and a tight coupling between them leads to higher performance without any significant loss in understanding the code. The derivation of these four classes from *DataStructure* is shown in Fig. 1.

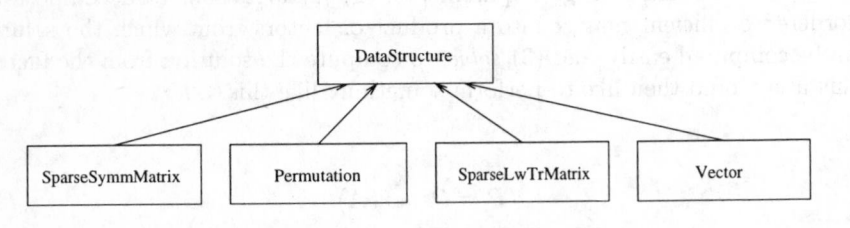

Fig. 1. High level structural classes

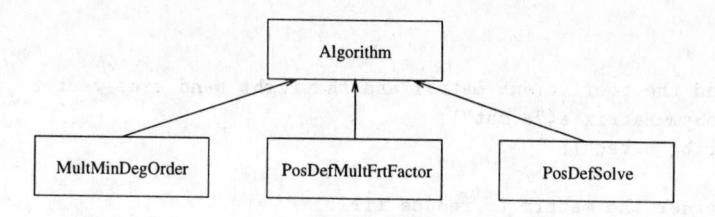

Fig. 2. Some high level algorithmic classes

At the same level the user also sees several algorithmic classes. First there are various ordering algorithms, such as *NestDissOrder* or *MultMinDegOrder*. Then there are factorization algorithms, like *PosDefLeftLookFactor*, *PosDefMultFrt-Factor* or *IndefMultFrtFactor*. Finally, the solve step can be performed by *Pos-DefSolve* or *IndefSolve* algorithms. Figure 2 describes the derivation of some of these classes from *Algorithm*. Using them one can easily write a solver (positive definite, for concreteness) shown in Fig. 3.

More details are available beyond this level of functionality. The factorization is split in two phases: *symbolic* and *numerical*. The symbolic factorization is guided by an elimination forest. The multifrontal method for numerical factorization uses an update stack and several frontal and update matrices, which are dense and symmetric. Pivoting strategies for indefinite systems can be controlled at the level of frontal and update matrices during the numerical factorization phase. Figures 4 and 5 depict the derivation of the corresponding structural and algorithmic classes.

Classes such as *SparseSymmMatrix*, *SparseLwTrMatrix*, and *Permutation* are implemented with multiple arrays of differing sizes. Several of these are arrays of indices that index into the other arrays, so that the validity of the state of a class depends on not only the individual internal arrays, but the interaction between several of them.

In a conventional sparse solver, these arrays are global and some of them are declared in different modules. A coefficient matrix, a factor, a permutation, or an elimination forest is not a well defined entity but the sum of scattered data. This inhibits software maintenance because of the tight coupling between disparate compilational units.

There are also significant benefits in terms of type safety. For instance, a permutation is often represented as an array of integers. It could be that the index of the old number holds the new position or vice versa. We use *oldToNew* and *newToOld* to refer to the two arrays. The problem is that interpreting a *newToOld* permutation as an *oldToNew* permutation yields a valid operation, though an incorrect permutation. It is easy for users to reverse these two, particularly when the names "permutation" and "inverse permutation" are applied since there is no agreement on whether *newToOld* is the former or the latter. Our *Permutation* class maintains both arrays internally and supplies each on demand.

```
main()
{
    /* Load the coefficient matrix and the right hand side vector. */
    SparseSymmMatrix a("a.mat");
    Vector b("b.vec");

    /* Reorder the matrix to reduce fill. */
    Permutation p(a.getSize());
    MultMinDegOrder order(a, p);
    order.run();

    /* Factor the reordered matrix. */
    SparseLwTrMatrix l(a.getSize());
    PosDefMultFrtFactor factor(a, p, l);
    factor.run();

    /* Declare algorithmic objects. */
    Vector x(a.getSize());
    PosDefSolve solve(l, p, b, x);
    solve.run();

    /* Save the solution. */
    x.save("x.vec");
}
```

Fig. 3. A direct solver for sparse, symmetric positive definite problems at the highest level

4 Design of the Lower Layers

While the larger part of our code deals with the design of the higher layers, most of the CPU time is actually spent in few computationally intensive loops. No advanced software paradigms are needed at this level so we concentrated on performance by carefully implementing these loops.

A major problem with C++ (also with C) is pointer aliasing, which makes code optimization more difficult for a compiler. We get around this problem by making local copies of simple variables in our kernel code. Another source of performance loss is complex numbers, since they are not a built-in in C++ data type as in Fortran. There is a template complex class in the Standard C++ library. Though this gives the compiler enough information to enforce all the rules as if it were a built-in datatype, it does not (indeed cannot) give the compiler any information about how to optimize for this class as if it were a built-in datatype.

We implemented our computationally intensive kernels both in C++ and Fortran 77. A choice between these kernels and between real and complex arithmetic can be made using compile-time switches. We defined our own class for complex numbers but we make minimal use of complex arithmetic operators,

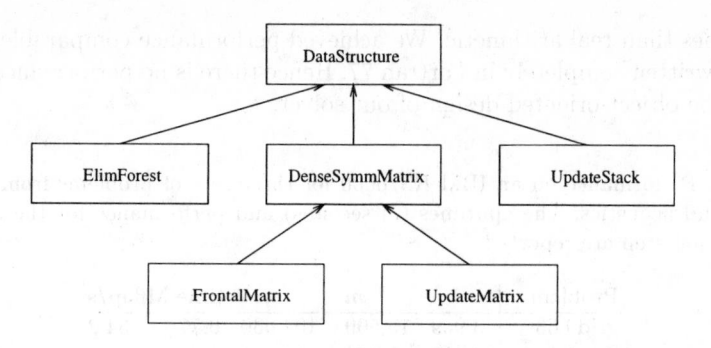

Fig. 4. Structural classes used by the multifrontal numerical factorization algorithms

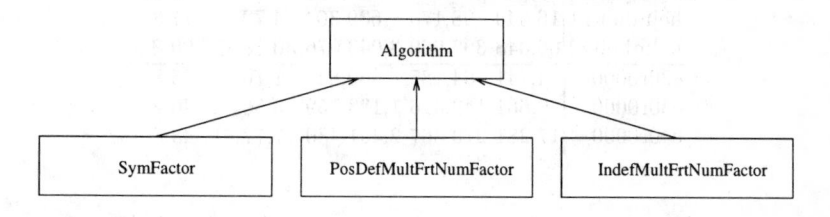

Fig. 5. Some symbolic and numerical factorization algorithmic classes

which are overloaded. The bulk of the computation is performed either in C++ kernels written in C-like style or in Fortran 77 kernels. Currently, we obtain better results with the Fortran 77 kernels.

5 Results

We report results obtained on a 66MHz IBM RS/6000 machine with 256 MB main memory, 128 KB L1 data cache and 2MB L2 cache, running AIX 4.2. Since this machine has two floating point functional units, each one capable of issuing one fused multiply-add instruction every cycle, its peak performance is theoretically 266 Mflop/s. We used the Fortran 77 kernels and we compiled the code with xlC 3.1.4 (-O3 -qarch=pwr2) and xlf 5.1 (-O4 -qarch=pwr2).

We show results for three types of problems: two-dimensional nine-point grids, Helmholtz problems, and Stokes problems, using multiple minimum degree ordering and multifrontal factorization. We use the following notation: n is the numbers of vertices in $G(A)$, (this is the order of the matrix), m is the number of edges in $G(A)$, and m^+ is the number of edges in $G^+(A, P)$, the filled graph. The difference between m^+ and m represents the fill. In Table 1 we describe each problem using these three numbers and we also provide the cputime and the performance for the numerical factorization step, generally the most expensive step of the computation. Higher performance is obtained for the Helmholtz problems because complex arithmetic leads to better use of registers and caches than real arithmetic. We achieved performance comparable to other solvers, written completely in Fortran 77. Hence there is no performance penalty due to the object-oriented design of our solver.

Table 1. Performance on an IBM RS/6000 for three sets of problems from fluid dynamics and acoustics. The cputimes (in seconds) and performance for the numerical factorization step are reported

Problem	n	m	m^+	time	Mflop/s
grid9.63	3,969	15,500	104,630	0.77	34.2
grid9.127	16,129	63,756	552,871	1.70	41.6
grid9.255	65,025	258,572	2,717,313	10.89	47.4
helmholtz0	4,224	24,512	130,500	0.77	62.3
helmholtz1	16,640	98,176	639,364	4.72	77.8
helmholtz2	66,048	392,960	3,043,076	30.88	90.8
e20r0000	4,241	64,185	369,843	1.70	35.8
e30r0000	9,661	149,416	1,133,759	6.56	40.2
e40r0000	17,281	270,367	2,451,480	17.77	43.6

We are currently implementing the solver in parallel using the message-passing paradigm. We plan to derive new classes to deal with the parallelism. Consider *FrontalMatrix* class, which stores the global indices in the *index* array and the numerical values in the *value* array. A *ParFrontalMatrix* class would need to add a *processor* array to store the owner of each column. A *ParUpdateMatrix* class may be derived in a similar way from *UpdateMatrix*. Some parallel algorithmic classes would be needed as well.

References

[1] S. Balay, W. D. Gropp, L. C. McInnes, and B. F. Smith. Efficient management of parallelism in object-oriented numerical software libraries. In *Modern Software Tools in Scientific Computing*. Birkhauser Press, 1997.

[2] A. M. Bruaset and H. P. Langtangen. Object-oriented design of preconditioned iterative methods in Diffpack. *ACM Trans. Math. Software*, pages 50–80, 1997.

[3] A. George and J. W. H. Liu. *Computer Solution of Large Sparse Positive Definite Systems*. Prentice Hall, 1981.

[4] A. Pothen and C. Sun. A distributed multifrontal algorithm using clique trees. Technical Report CS-91-24, Computer Science, Penn State, Aug 1991.

[5] C. Ashcraft, J. Lewis, and R. Grimes. Accurate symmetric indefinite linear equation solvers. Preprint, Boeing Information Sciences, 1995. To appear in SIAM J. Matrix Analysis and its Applications.

[6] E. Arge, A. M. Bruaset, and H. P. Langtangen. Object-oriented numerics. In *Numerical Methods and Software Tools in Industrial Mathematics*, pages 7–26. Birkhauser, 1997.

[7] G. Booch. *Object-Oriented Analysis and Design with Applications*. Benjamin Cummings Publishing Company, 1994. Second edition.

Janus: A C++ Template Library for Parallel Dynamic Mesh Applications

Jens Gerlach, Mitsuhisa Sato, and Yutaka Ishikawa

Tsukuba Research Center of the Real World Computing Partnership, Tsukuba, Japan

Abstract. We propose *Janus*, a C++ template library of container classes and communication primitives for parallel dynamic mesh applications. The paper focuses on two-phase containers that are a central component of the Janus framework. These containers are quasi-constant, i.e., they have an extended initialization phase after which they provide read-only access to their elements. Two-phase containers are useful for the efficient and easy-to-use representation of finite element meshes and generating sparse matrices. Using such containers makes it easy to encapsulate irregular communication patterns that occur when running finite element programs in parallel.

1 Introduction

If we think of a finite element program as a collection of related objects on which operations are performed, we recognize that there are two types of application objects. The first are sets that represent spatial structures, and the second are numerical functions on these sets. Here are some examples of spatial structures and functions on them.

- Given the node set N, many physically relevant data are represented by functions $f : N \mapsto \mathbb{R}$.
- The element matrices are a function from the triangulation $e : T \mapsto \mathbb{R}^{p \times p}$, where p denotes the number of degrees of freedom per element.
- The system matrix m is a function $m : \mathcal{N} \mapsto \mathbb{R}$, where $\mathcal{N}$ is a subset of $N \times N$ that represents the sparsity pattern of the matrix.

The parallelism of finite element methods is mainly data parallelism with respect to the meshes. Using parallel computers with a distributed memory architecture requires therefore a partition of the triangulation T and the node set N. Whatever communication occurs when running a finite element program in parallel, it is caused by a relation of the meshes. A problem hereby is that due to their irregularity and size the relations must be partitioned themselves.

The driving motivation behind the design of the Janus framework is to provide application-oriented, easy-to-use, efficient abstractions for the fundamental components of finite elements methods mentioned above. Janus offers building blocks to represent (possibly partitioned) spatial structures and functions on them. Moreover communication must be expressed explicitly based on the mesh relations.

Janus is implemented as a C++ template library. The library is generic with respect to the numerical types, the way the user wants to represent mesh points, and the mapping information that is used for partitioning the meshes.

A fundamental concept in Janus is that of *two-phase containers* which are used to represent spatial structures with a non-trivial initialization phase. The lifetime of a two-phase container is separated into a *generation phase* and an *access phase*. The transition from one to the other phase is marked by a call to its `freeze` method. Such a concept is useful for the implementation of finite element methods, for three reasons. First, an adaptive finite element method can be considered as a succession of *generation* and *computation* phases (cf. [1]). The same holds for the underlying patterns of sparse matrices. Second, the necessary initializations are usually too complex to be performed in one call of a C++ constructor. Extending the initialization phase helps to make the change of data structures transparent to the application programmer. Third, communication that might occur while generating distributed meshes can be delayed until the call of the `freeze` method.

Another important concept is that of an *associated container* that is used for the representation of numerical functions on (distributed) spatial structures.

An overview of these containers and their use in a sequential context is presented in Sect. 2. Aspects of the parallel implementation and optimizations that are enabled by the use of two-phase containers are discussed in Sect. 3. We explain how the use of two-phase containers allows one to analyze irregular communication patterns as they occur in finite element sparse matrices. This is an important optimization for the iterative solution of parallel finite element problems.

2 Concepts and Classes in Janus

Both from a conceptional and implementation point of view Janus is based on the containers and algorithms of the Standard C++ Library [2], also known as the Standard Template Library (STL) [3]). The STL is not only a collection of fundamental data structures, generic classes, and algorithms. It defines *concepts*, i.e. generic sets of type requirements and its container classes are models of these concepts, i.e., they are types that satisfy these requirements. The idea is that: "Using concepts makes it possible to write programs that cleanly separate interface from implementation" [3].

2.1 Two-Phase Containers

A *two-phase container* is a variable-sized container that supports insertion of elements. However, all insertions must have been finished *before* any element of the container can be *accessed*. Only *non-mutating* access is allowed.

The phase in which insert operations are allowed is called the *generation phase* or *first phase*. The phase in which access is allowed is called *access phase* or *second phase*. The transition from the first to the second phase is marked by a call to the `void freeze()` method of a two-phase container.

Containers that follow these requirements represent application objects that have a non-trivial yet clearly distinguishable initialization procedure. Typical examples are finite element meshes or sparse matrix patterns whose structure is not known at compile time.

A two-phase container can be "frozen" only once and it has no `thaw` method that would allow new insertions. This means that a two-phase container cannot be used to implement meshes that are meant to be modified after their initialization. However, this is only an apparent restriction, since from a conceptional point of view it is often easier to represent mesh modification by creating a new mesh out of an existing one (cf. [1]).

The `FixedSet` Container Family This provides two template classes that are models of the two-phase container concept, namely the `OrderedFixedSet` and the `HashedFixedSet` templates. Elements of each container type must be unique.

The main difference between these two classes is that the `OrderedFixedSet` uses the STL `set` template class to initially store its data, whereas the other is implemented by the STL `hash_set` container. Both containers provide read-only random access to their elements. This is very natural, since in the second phase (when the container is frozen), it is no problem to number its elements from 0 to `size()`. This property of two-phase containers can be exploited for a very efficient implementation of vector classes for finite element methods, which is explained in Sect. 2.2.

Note that the actual details of the representation of the sets (red-black tree or hash table) are hidden from the user. It is very easy to switch between both implementation strategies or even to mix them. This is in contrast to implementation strategies that expose such low-level details to the application programmer [4].

Use of Two-Phase Containers The code fragment in Fig. 1 shows a typical use of a two-phase container. Given the triangles of a finite element mesh, its nodes (in this particular case the vertices of the triangles) will be generated. We use a six-tuple of integers to denote triangles and their nodes. This allows us to express the triangle-vertex relation by simple index arithmetic [5, 6]. To get the vertices of a triangle on a certain level of an adaptively refined mesh the short inline function `vertices` must be called. This is done for each triangle and the resulting nodes are inserted in the node set `nodes`. Even if the same node is inserted several times the implementation of the container assures that it occurs only once.

After all vertices have been inserted, the `nodes` container is frozen. In case of an `OrderedFixedSet` container its elements are copied from a STL-`set` container

```
typedef OrderedFixedSet<Index<6>,less<Index<6> > > Triangles;
typedef OrderedFixedSet<Index<6>,less<Index<6> > > Nodes;

Nodes create_nodes(int level, const Triangles& triangles)  {
  Nodes nodes;
  Triangles::const_iterator i;
  for(i = triangles.begin(); i != triangles.end();i++)  {
    Tuple<Index<6>,3> v = vertices(*i, level);
    for(size_t j = 1; j <= 3; j++) nodes.insert(v[j]);
  }
  nodes.freeze();
  return nodes;
}
```

Fig. 1. Using a two-phase container for the representation of the nodes of a finite element mesh

that was used during the initialization phase to a dynamically allocated fixed-size array represented by the STL-`vector` container.

The `FixedRelation` Container Family This family consists of two-phase containers to represent relations between two sets. Therefore they store pairs of elements of other sets. Except for some additional methods and type information about the underlying sets the interface of these classes is the same as for `FixedSet` containers.

There is a special member of this family called `IndexedFixedRelation`. When calling `freeze()` the position of the components of its pairs with respect to the underlying sets are determined. It is shown in Sect. 2.3 how this can be used for the efficient implementation of sequential finite element methods.

2.2 Associated Containers

Associated containers are primarily used for the efficient representation of numerical functions on sets represented by two-phase containers. An associated container is, by definition, a random-access container whose size is determined by that of a another container that represents the underlying set. When an associated container is initialized it gets a reference to its underlying set object, which must be a fixed-size container or a frozen two-phase container. This allows for efficient storage of the elements, for example the STL `valarray<T>` could be used. Elements of associated containers can be accessed by random access or by access through elements of the underlying set (the `at` method).

The `SetArray` class template is Janus' standard model of an associated container. It offers no direct support for numerical operations. These services are provided by the template classes `SetVector` and `SetMatrix` which are wrappers around `SetArray`. The main difference between both containers is that

SetMatrix requires that the underlying set is a member of the FixedRelation container family.

2.3 Interaction of Two-Phase and Associated Containers

For each triangle the average value of a grid function u on the index set nodes will be computed and stored in a grid function x on the index set triangles. To determine the vertices of a triangle we use the vertices method again. Note that we iterate over the triangles through random access to x.set() which returns a reference to triangles.

```
void average(int level, const SetArray<Nodes,double>& u,
             SetArray<Triangles,double>& x) {
  for(size_t i = 0; i < x.size(); i++) {
    Index<6> triangle = x.set()[i];
    Tuple<Index<6>,3> v = vertices(triangle, level);
    x[i] = (u.at(v[1]) + u.at(v[2]) + u.at(v[3])) / 3.0;
  }
}
```

Fig. 2. Implementation of the function average

The implementation shown in Fig. 2 looks quite appealing, but there are two problems with this usage of the at method. First, the at method will not work in the parallel case because the data that it tries to access may reside in another computational domain, and Janus does not support (for performance reasons) remote accesses to *individual* elements. Second, the overhead may be large even in the sequential case since a call of at causes a non-trivial search in the underlying set.

A solution to both problems is to compute in advance the relation between the triangles and their vertices and to store them in a variable of type Tuple<Triangles_Nodes,3>. That is, we consider the triangle triangle-vertex relation as three separate relations.

Note that in the example in Fig. 3 the template IndexedFixedRelation (mentioned in Sect. 2.1) is used. The precalculated positions can be accessed through the methods index1(size_t) and index2(size_t). This means that in the sequential case the average procedure can be implemented as in Fig. 4.

Note that this use of the precalculated indices is nothing more than the traditional "index arrays" that are typically used in Fortran programs. In Janus these helper objects are set up when the container that holds relation is frozen. This computation is therefore transparent to the user.

```
typedef IndexedFixedRelation<Triangles,Nodes> Triangles_Nodes;

Tuple<Triangles_Nodes,3>
triangle_vertex(const Triangles& t, const Nodes& n, int level) {
  Tuple<Triangles_Nodes,3> result(Triangles_Nodes(t,n));
  for(Triangles::const_iterator i=t.begin(); i!= t.end(); i++) {
    Index<6>              triangle = *i;
    Tuple<Index<6>,3> node = vertices(triangle,level);
    for(size_t j = 1; j <= 3; j++)
      result[j].insert(make_pair(triangle,node[j]));
  }
  for(size_t j = 1; j <= 3; j++)  result[j].freeze();
  return result;
}
```

Fig. 3. Creation of the triangle vertex relation

```
void average(const SetArray<Nodes,double>& u,
const Tuple<Triangles_Nodes,3>& r, SetArray<Triangles,double>& x) {
  for(size_t i = 0; i < x.size(); i++)
    x[i] = (u[r[1].index2(i)] + u[r[2].index2(i)] +
            u[r[3].index2(i)]) / 3.0;
}
```

Fig. 4. Revised sequential implementation of the function **average**.

3 Parallel Environment

With respect to a parallel implementation the programmer should have an SPMD programming model in mind. The library supports expressing data parallelism on the level of meshes. This requires first of all that programmers have a good model to represent mesh nodes and elements. We advocate representations of meshes by so-called index spaces, sets of integer tuples [5, 6, 1].

The great advantage of our indexing technique is that it provides application-oriented global names that are independent from implementation details. This allows expression of communication relations independent from the mapping of the indices onto the underlying hardware architecture. The approach of using integer tuples to place and retrieve data recalls the concept of tuple spaces in Linda [7]. However, in Janus these integer tuples are stored in two-phase containers whose access semantics are formed after the usage cycle of finite element meshes. This allows locally fast random access to the data.

3.1 Mapped Containers

In a parallel and distributed environment the finite element meshes must be distributed over a group of abstract processes, which are called *domains* in Janus. As in MPI these processes are denoted by integers [8].

To represent distributed meshes in Janus the programmer uses *mapped* (two-phase) containers. Mapped containers have an additional template parameter that serves as a mapping type. As a mapping type any class that has a method domain can be used that assigns an integer to its argument. The mapped container uses the mapping type to decide to which domain an object that is inserted will be mapped. The idea of using mapping type template parameters has been taken from the runtime library of the PROMOTER programming model [9, 10]. It gives the user greater flexibility in choosing appropriate mapping strategies.

If an object is inserted into a mapped container then the mapping type is taken to check to which domain the object belongs. If the domain is the same as the one of the mapped container then it is inserted locally. Otherwise, it is put in a temporary buffer. When calling the freeze methods of the mapped container, the temporary buffers are sent to the appropriate domains where the objects are inserted. Delaying the communication is possible since elements are accessed only after the freeze method has been called.

3.2 Communication in Janus

To express communication in Janus the user must explicitly describe which points belong to the underlying mesh relation. Figure 3 showed the example of creating the triangle vertex relation.

Note that in Janus the user describes the relation on the level of mesh points, in an application-oriented way. When creating a relation the user need not specify where the mesh points he refers to are actually stored. This necessary information is obtained by the library from the mapping objects of the mapped two-phase containers.

Since the relation itself is stored in a two-phase container it will not change during its use. Thus it can be examined before its first use. Analyzing the sparsity patterns allows message buffers of the right size to be created in advance, thus reducing the communication overhead. This is a very important optimization for parallel sparse matrix multiplication, which is a key component of iterative methods. They are the preferred methods for the solution of large scale finite element problems.

4 Concluding Remarks

We have presented the major concepts of a template library for data parallel adaptive mesh applications. The concept of a two-phase container provides simple yet sufficient and efficient support for irregular structures such as finite element meshes and sparse matrix patterns. Two-phase containers are beneficial in a sequential and parallel context and serve as a useful base for other concepts such as associated containers. Using two-phase containers for the description of mesh relations allows irregular communication patterns to be analyzed when they are created.

Currently we use a prototype of Janus for the parallel finite element analysis on two-dimensional meshes. For the solution of the linear systems we use the conjugate gradient method with a simple diagonal preconditioner. In future we will incorporate multilevel preconditioners and adaptive refinement into the solver. The necessary abstractions are already contained in Janus.

References

[1] J. Gerlach. Application of natural indexing to adaptive multilevel methods for linear triangular elements. TR 97-010, Technical Report of the Real World Computing Partnership, Japan http://www.rwcp.or.jp/people/jens/publications/TR-97-010.

[2] Bjarne Stroustrup. *The C++ Programming Language, Third Edition*. Addison-Wesley, 1997.

[3] *Standard Template Library Programmer's Guide*, http://www.sgi.com/Technology/STL/.

[4] M. Griebel, G. Zumbusch. Hash-storage techniques for adaptive multilevel solvers and their domain decomposition parallelization. *Contemporary Mathematics* Vol. 218, pp. 279-286.

[5] J. Gerlach, M. Sato, Y. Ishikawa. A framework for parallel adaptive finite element methods and its template based implementation in C++, Proc. 1st Int. Conf. on Scientific Computing in Object-Oriented Parallel Environments, Marina del Rey, CA (1997), Lecture Notes in Computer Science, LNCS 1343, Springer Verlag, 1997.

[6] J. Gerlach, G. Heber. Fundamentals of natural indexing for simplex finite elements in two and three dimensions. TR 97-008, Technical Report of the Real World Computing Partnership, Japan http://www.rwcp.or.jp/people/jens/publications/TR-97-008

[7] David Gelernter. Generative communication in Linda. *ACM Trans. on Programming Languages and Systems* Vol. 2, No. 1, pp. 80–112, January 1985.

[8] W. Gropp, E. Lusk, A. Skjellum. *Using MPI*, The MIT Press, 1994.

[9] Giloi W.K., Kessler M., Schramm, A. PROMOTER: A high level, object-parallel programming language. Proc. Int. Conf. on High Performance Computing, New Dehli, India, December 1995.

[10] Bi Hua. Object-oriented data parallel programming in C++. Proc. PDPTA'97, Las Vegas, USA, CSREA 1997, RWC-D-97-015.

Arrays in Blitz++

Todd L. Veldhuizen

Indiana University, Bloomington, IN, USA

Abstract. The Blitz++ library provides numeric arrays for C++ with efficiency that rivals Fortran, without any language extensions. Blitz++ has features unavailable in Fortran 90/95, such as arbitrary transpose operations, array renaming, tensor notation, partial reductions, multi-component arrays and stencil operators. The library handles parsing and analysis of array expressions on its own using the expression templates technique, and performs optimizations (such as loop transformations) which have until now been the responsibility of compilers.

1 Introduction

The goal of the Blitz++ library is to provide a solid "base environment" of arrays, matrices and vectors for scientific computing in C++. This paper focuses on arrays in Blitz++, which provide performance competitive with Fortran and superior functionality. The design of Blitz++ has been influenced by Fortran 90, High-Performance Fortran, the Math.h++ library [1], A++/P++ [2], and POOMA [3]. It incorporates various features from these environments, and adds many of its own. This paper concentrates on the unique features of Blitz++ arrays.

2 Overview

Multidimensional arrays in Blitz++ are provided by the class template `Array<T, N>`. The template parameter `T` is the numeric type stored in the array, and `N` is its rank. This class supports a variety of array models:

- Arrays of scalar types, such as `Array<int,2>` and `Array<float,3>`
- Complex arrays, such as `Array<complex<float>,2>`
- Arrays of user-defined types. For example, if `Polynomial` is a class defined by the user (or another library), `Array<Polynomial,2>` is a two dimensional array of `Polynomial` objects.
- Nested homogeneous arrays using the Blitz++ classes `TinyVector` and `TinyMatrix`. For example, `Array<TinyVector<float,3>,3>` is a three-dimensional vector field.
- Nested heterogeneous arrays, such as `Array<Array<int,1>,1>`, in which each element is an array of variable length.

2.1 Storage Layout and Reference Counting

Array objects are lightweight views of a separately allocated data block. This design permits a single block of data to be represented by several array views [1]. Each array object contains a *descriptor* (also called a *dope vector*) which specifies the memory layout. The descriptor contains a pointer to the array data, lower bounds for the indices, a shape vector, a stride vector, reversal flags, and a *storage ordering* vector. This last is a permutation of the dimension numbers $[1, 2, \ldots, N]$ which indicates the order in which dimensions are stored in memory. Fortran-style column-major arrays correspond to $[1, 2, \ldots, N]$, and C-style row-major arrays correspond to $[N, N-1, \ldots, 1]$. Reversal flags indicate whether each dimension is stored in ascending or descending order.

The storage ordering vector and reversal flags allow arrays to be stored in any one of $N!2^N$ orderings. Only two of these – C and Fortran-style arrays – are frequently used. There are occasional uses for other orderings: some image formats store rows from bottom to top, which can be handled transparently by a reversal flag.

Arrays are reference-counted: the number of arrays referencing a data block is monitored, and when no arrays refer to a data block it is deallocated. Reference counting provides the benefits of garbage collection, and allows functions to return array objects efficiently:

```
Array<float,2> someUserFunction(Array<float,2>&);
```

Reference-counting and flexible storage formats support useful $O(1)$ array operations:

- *Arbitrary transpose operations:* The dimensions of an array can be permuted using the `transpose(...)` member function. This code makes B a shared view of A, but with the first and second dimensions swapped:

```
Array<float,3> A(3,3,3);      // A 3x3x3 array
Array<float,3> B=A.transpose(secondDim,firstDim,thirdDim);
```

The integer constants `firstDim`, `secondDim`, ... are intended to improve readability, and hide confusion over whether the first dimension is 1 (as in Fortran) or 0 (as in C).
- *Dimension reversals:* Each dimension can be independently reversed. If A contains a two-dimensional colour image, then

```
Array<RGB24,2> B = A.reverse(firstDim);
```

flips the image vertically.
- *Array relabelling:* Since array objects are really lightweight handles, arrays can be swapped and relabelled in constant time. This is very useful in time-stepping PDEs. If A1, A2 and A3 represent a field at three consecutive timesteps, `cycleArrays(A1,A2,A3)` relabels the arrays for the next time step: $[A1,A2,A3] \leftarrow [A2,A3,A1]$. This avoids costly copying of the array data.
- *Array interlacing:* Blitz++ allows arrays of the same shape to be interlaced in memory. Such an arrangement improves data locality, which can increase performance in some situations.

2.2 Subarrays and Slicing

Subarrays in Blitz++ are fully functional `Array` objects which have a shared view of the array data. Subarrays can either be full rank or lesser rank. Blitz++ supplies `Range` objects which emulate the Fortran 90 range syntax. Any combination of `Range` and integer values can be used to obtain a subarray:

```
Array<float,3> A(64,64,64);   // A 64x64x64 array

// C refers to the 2D slice A(10..63, 15, 0..63)
Array<float,2> C = A(Range(10,toEnd), 15, Range::all());
Array<float,1> D = A(Range(fromStart,30),15,20); //A(0..30,15,20)
```

The use of `fromStart` and `toEnd` is after [1]. An optional third parameter to the `Range` constructor specifies a stride, so subarrays do not have to be contiguous.

3 Array Expressions

Array expressions in Blitz++ are implemented using the expression templates technique [4]. Prior to expression templates, use of overloaded operators meant generating temporary arrays, which caused huge performance losses. In Blitz++, temporary arrays are never created. Since its original development, the expression templates technique has grown substantially more complex and powerful [5, 6]. Its present incarnation in Blitz++ supports a wide variety of useful notations and optimizations. The next sections overview the main features of the Blitz++ expression templates implementation from a user perspective.

3.1 Operators

Any operator which is meaningful for the array elements can be applied to arrays. For example:

```
    Array<float,2> A, B, C, D;  // ...
    A = B + (C * D);

    Array<int,1> E, F, G, H;    // ...
    E |= (F & G) >> H;
```

Operators are always applied in an elementwise manner. Users can create arrays of their own classes, and use whichever overloaded operators they have provided:

```
    class Polynomial {
      // define operators + and *
    };

    Array<Polynomial,2> A, B, C, D;  // ...
    A = B + (C*D);   // results in appropriate calls
                     // to Polynomial operators
```

Math functions provided by the standard C++, IEEE and System V math libraries may be used on arrays, for example sin(A) and lgamma(B).

Arrays with different storage formats can appear in the same expression; for example, a user can add a C-style array to a Fortran array. Blitz++ transparently corrects for the storage formats. Blitz++ allows arrays of different numeric types to be mixed in an expression. Type promotion follows the standard C rules, with some modifications to handle complex numbers and user-defined types.

Blitz++ supplies a set of *index placeholder* objects which allow array indices to be used in expressions. This code creates a Hilbert matrix:

```
Array<float,2> A(4,4);

// i and j are index placeholders
firstIndex i;
secondIndex j;

A = 1.0 / (1+i+j); // Sets A(i,j) = . . . for all (i,j)
```

3.2 Tensor Notation

Blitz++ provides a notation modelled on tensors. Here is an example of mathematical tensor notation:

$$C^{ijk} = A^{ij}x^k - A^{jk}y^i$$

In Blitz++, this equation can be coded as:

```
using namespace blitz::tensor;

C = A(i,j) * x(k) - A(j,k) * y(i);
```

The tensor indices i,j,k,... are special objects concealed in the namespace blitz::tensor. Users are free to declare their own tensor indices with different names if they prefer. Tensor indices specify how arrays are oriented in the domain of the array receiving the expression (Fig. 1). Any missing tensor indices are interpreted as spread operations; for example, the A(i,j) term in the above example is spread over the k index.

Unlike real tensor notation, repeated indices do not imply contraction. For example, the tensor expression $C^{ij} = A^{ik}B^{kj}$ implies a summation over k. In Blitz++, contractions must be written explicitly using a partial reduction (described later):

```
Array<float,2> A, B, C;  // ...
C(i,j) = sum(A(i,k) * B(k,j), k);
```

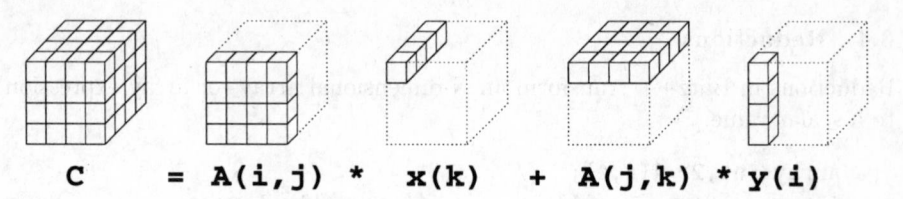

Fig. 1. Illustration of Blitz++ tensor notation: the indices specify how arrays are oriented in the domain of the array receiving the result

3.3 Stencil Objects and Operators

Blitz++ provides a stencil object mechanism that removes much of the drudgery from writing finite difference equations. One of the Blitz++ example programs is a three-dimensional computational fluid dynamics simulation. In each iteration, the velocity field is time-stepped according to the equation

$$\mathbf{V} \leftarrow \mathbf{V} + \Delta t \left(\rho^{-1} \left(\eta \nabla^2 \mathbf{V} - \nabla P + \mathbf{F} \right) - \mathbf{A} \right)$$

where $\mathbf{V}$, P, $\mathbf{F}$ and $\mathbf{A}$ are velocity, pressure, force, and advection. Implementing this equation using 4th-order accurate finite differencing in Fortran requires a set of mammoth equations with approximately 70 terms. Using a Blitz++ stencil object, the equation is written as:

```
nextV = *V + delta_t * (recip_rho * (eta * Laplacian3DVec4(V,
    geom) - grad3D4(P, geom) + *force) - *advect);
```

The vector fields V, force and advect are implemented as arrays of 3-vectors. This eliminates the need to represent each vector field as three separate arrays, common in Fortran implementations. The *stencil operators* Laplacian3DVec4 and grad3D4 are provided by Blitz++, and implement 4th-order Laplacian and gradient operators. The Laplacian3DVec4 operator expands into a 45-point stencil. Blitz++ supplies stencil operators for forward, central and backward differences of various orders and accuracies; built on top of these are divergence, gradient, curl, mixed partial, and Laplacian operators.

Blitz++ provides special support for vector fields (and in general, multi-component/multispectral arrays). The [] operator is overloaded for easy access to individual components of a multicomponent array. For example, this code initializes the force field with gravity:

```
const int x = 0, y = 1, z = 2;
force[x] = 0.0;
force[y] = 0.0;
force[z] = gravity;
```

3.4 Reductions

Reductions in Blitz++ transform an N-dimensional array (or array expression) to a scalar value:

```
Array<int,2> A(4,4);   // ...
int result1 = sum(A);          // sum all elements
int result2 = count(A == 0);   // count zero elements
```

Available reductions are sum, product, min, max, count, minIndex, maxIndex, any and all. *Partial* reductions transform an N-dimensional array (or array expression) to an N-1 dimensional array expression. The reduction is performed along a single rank:

```
Array<int,2> A(2,4);
Array<int,1> B(2);

A = 0, 1, 1, 5,
    3, 0, 0, 0;

B = sum(A, j);       // Reduce along rows: B = [ 7   3 ]
```

Reductions can be chained: for example, this code finds the row with the minimum sum of squares:

```
Array<float,2> A(N,N);   // ...
int minRow = minIndex(sum(pow2(A),k));
```

4 Optimizations

The expression tempaltes technique allows Blitz++ to parse array expressions and generate customized evaluation kernels at compile time. To achieve good performance, Blitz++ performs many loop transformations which have traditionally been the responsibility of optimizing compilers:

- *Loop interchange and reversal:* Consider this bit of code, which is a naïve implementation of the array operation A = B + C:

```
for (int i=0; i < N1; ++i)
  for (int j=0; j < N2; ++j)
    for (int k=0; k < N3; ++k)
      A(i,j,k) = B(i,j,k) + C(i,j,k);
```

The layout of these arrays in memory is unknown at compile time. If the arrays are stored in column-major order, this code will be very inefficient because of poor data locality. For large arrays, an entire cache line would have to be loaded for each element access. To avoid this problem, Blitz++ selects a traversal order at run-time such that the arrays are traversed in memory-storage order.

- *Hoisting stride calculations:* The inner loop of the above code fragment would expand to contain many stride calculations. Blitz++ generates code which hoists the invariant portion of the stride arithmetic out of the innermost loop.
- *Collapsing inner loops:* Suppose that in the above code fragment, N3 is quite small. Loop overhead and pipeline effects will conspire to cause poor performance. The solution is to convert the three nested loops into a single loop. At runtime, Blitz++ collapses the inner loops whenever possible.
- *Partial unrolling:* Many compilers partially unroll inner loops to expose low-level parallelism. For compilers that won't, Blitz++ does this unrolling itself.
- *Common stride optimizations:* Blitz++ tests at run-time to see if all the arrays in the expression have a unit or common stride. If so, faster evaluation kernels are used.
- *Tiling:* Blitz++ detects the presence of stencils, and does tiling to ensure good cache use.

4.1 Benchmark Results

Figure 2 shows performance of the Blitz++ classes `Array` and `Vector` for a DAXPY operation on the Cray T3E-900 (single PE) using KAI C++. The Blitz++ classes achieve the same performance as Fortran 90.[1] The native BLAS library is able to outperform both Fortran 90 and Blitz++.[2] Without expression templates, performance is typically 30% that of Fortran.

Table 1 shows performance of Blitz++ arrays on 21 loop kernels used by IBM for benchmarking the RS/6000. Performance is reported as a fraction of Fortran performance: > 100 is faster, and < 100 is slower. The fastest native Fortran compiler was used, with typical optimization switches (-O3, -Ofast). The loop kernels and makefiles are available as part of the Blitz++ distribution.

Table 1. Performance of Blitz++ on 21 loop kernels, relative to Fortran

Platform/ Compiler	Out of cache		In-cache (peak)	
	Median	Mean	Median	Mean
Cray T3E/KCC	95.7%	86.4%	98.1%	88.4%
HPC-160/KCC	100.2%	97.5%	95.1%	93.4%
Origin 2000/KCC	88.1%	87.3%	79.8%	78.6%
Pentium II/egcs	98.4%	98.5%	79.6%	82.6%
RS 6000/KCC	93.5%	90.7%	97.3%	93.2%
UltraSPARC/KCC	91.1%	86.8%	79.0%	78.3%

---

[1] Fortran 77 is no longer supported on the T3E, and is actually slower. The flags used for the f90 compiler were -O3,aggress,unroll2,pipeline3.

[2] Although not yet implemented, it is possible to do pattern matching to native BLAS using expression templates, an idea due to Roldan Pozo.

230

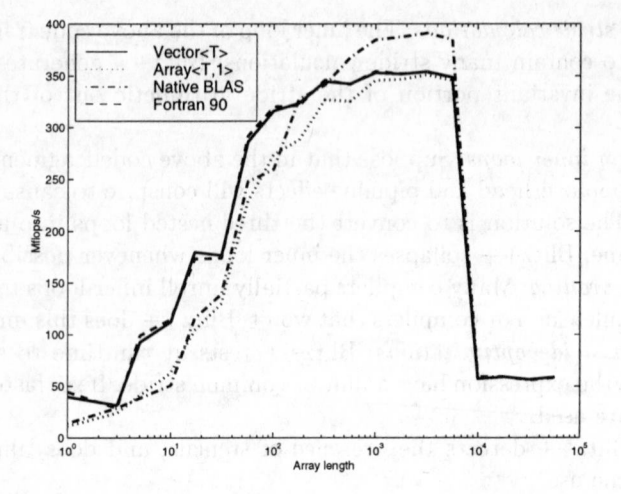

Fig. 2. DAXPY benchmark on the Cray T3E (single PE)

Acknowledgments This work was supported in part by NSERC (Canada), and by the Director, Office of Computational and Technology Research, Division of Mathematical, Information, and Computational Sciences of the U.S. Department of Energy under contract number DE-AC03-76SF00098. This research used resources of NERSC and the Advanced Computing Laboratory (LANL) which are supported by the Office of Energy Research of the U.S. Department of Energy, and of ZAM (Research Centre Jülich, Germany).

References

[1] Thomas Keffer and Allan Vermeulen. *Math.h++ Introduction and Reference Manual*. Rogue Wave Software, Corvallis, Oregon, 1989.

[2] Rebecca Parsons and Daniel Quinlan. A++/P++ array classes for architecture independent finite difference computations. In *Proc. 2nd Annual Object-Oriented Numerics Conf. (OON-SKI'94)*, pages 408–418, April 24–27, 1994.

[3] John V. W. Reynders, Paul J. Hinker, Julian C. Cummings, Susan R. Atlas, Subhankar Banerjee, William F. Humphrey, Steve R. Karmesin, Katarzyna Keahey, M. Srikant, and Marydell Tholburn. POOMA. In Gregory V. Wilson and Paul Lu, editors, *Parallel Programming Using C++*. MIT Press, 1996.

[4] Todd L. Veldhuizen. Expression templates. *C++ Report*, 7(5):26–31, June 1995. Reprinted in *C++ Gems*, ed. Stanley Lippman.

[5] Scott W. Haney. Beating the abstraction penalty in C++ using expression templates. *Computers in Physics*, 10(6):552–557, Nov/Dec 1996.

[6] Geoffrey Furnish. Disambiguated glommable expression templates. *Computers in Physics*, 11(3):263–269, May/June 1997.

Array Design and Expression Evaluation in POOMA II*

Steve Karmesin, James Crotinger, Julian Cummings, Scott Haney, William Humphrey, John Reynders, Stephen Smith, and Timothy J. Williams

Los Alamos National Laboratory, Los Alamos, NM USA

Abstract. POOMA is a templated C++ class library for use in the development of large-scale scientific simulations on serial and parallel computers. POOMA II is a new design and implementation of POOMA intended to add richer capabilities and greater flexibility to the framework. The new design employs a generic **Array** class that acts as an interface to, or *view* on, a wide variety of data representation objects referred to as *engines*. This design separates the interface and the representation of multidimensional arrays. The separation is achieved using compile-time techniques rather than virtual functions, and thus code efficiency is maintained. POOMA II uses PETE, the Portable Expression Template Engine, to efficiently represent complex mathematical expressions involving arrays and other objects. The representation of expressions is kept separate from expression evaluation, allowing the use of multiple evaluator mechanisms that can support nested where-block constructs, hardware-specific optimizations and different run-time environments.

1 Introduction

Scientific software developers have struggled with the need to express mathematical abstractions in an elegant and maintainable way without sacrificing performance. The POOMA (Parallel Object-Oriented Methods and Applications) framework [1, 2], written in ANSI/ISO C++, has demonstrated both high expressiveness and high performance for large-scale scientific applications on platforms ranging from workstations to massively parallel supercomputers. POOMA provides high-level abstractions for multidimensional arrays, physical meshes, mathematical fields, and sets of particles. POOMA also exploits techniques such as expression templates [3] to optimize serial performance while encapsulating the details of parallel communication and supporting block-based data compression. Consequently, scientists can quickly assemble parallel simulation codes by focusing directly on the physical abstractions relevant to the system under study and not the technical difficulties of parallel communication and machine-specific optimization.

POOMA II is a complete rewrite of POOMA intended to further increase expressiveness and performance. The array and field concepts have been redesigned

* This work was performed under the auspices of the U.S. Department of Energy by Los Alamos National Laboratory under Contract No. W-7405-Eng-36.

to use a powerful and flexible view-based architecture that decouples interface and representation. Expressions involving arrays and fields are packaged and manipulated using an enhanced version of PETE, the Portable Expression Template Engine. These expressions can operate on subsets of the data, specified via multiple-dimensional domain objects. Finally, the expressions are efficiently evaluated by evaluator objects. These evaluators support a variety of run-time systems, ranging from immediate serial evaluation to thread-based parallel evaluation, as well as complex constructs like where-blocks.

2 Arrays and Engines

An array is a logically rectilinear, N-dimensional table of numeric elements. Most array implementations store their data in a contiguous block of memory and apply Fortran or C conventions for interpreting this data as a multidimensional array. Unfortunately, these two storage conventions do not span the full range of array types encountered in scientific computing: diagonal, banded, symmetric, sparse, etc. One can even imagine arrays that use no storage, computing their element values as functions of their indices or via expressions involving other arrays. One approach to dealing with differing array storage strategies is to simply create new array classes for each case: `BandedArray`, `SparseArray`, and so on. However, this is wasteful since all of these variants have very similar interfaces.

POOMA II's array class provides a uniform interface independent of how the data is stored or computed, without incurring the overhead of C++ virtual function calls. This is accomplished by introducing the concept of an *engine*. An engine is an object that provides a common interface for randomly accessing and changing elements without the need for the user of the engine to know how the elements are stored. For example, an engine that manages a 100×200 "brick" of double-precision values is declared as:

```
Engine<2, double, Brick> brick(100, 200);
```

The domain of this engine is the tensor product of $[0 \ldots 99]$ by $[0 \ldots 199]$. Similarly, an engine that manages a brick of data distributed across a parallel machine in a manner specified by an object `layout` is declared as:

```
Engine<2, double, Distributed> dbrick(100, 200, layout);
```

The domain and range of `dbrick` are identical to that of `brick`, as is the interface for accessing elements. However, the implementations are quite different.

Note that engine classes are all specializations of a common template, `Engine`. A tag is used to specify a particular engine, such as `Brick` or `Distrubuted`, allowing useful default template parameters to be chosen for the array class.

Engines represent a low-level abstraction: getting single elements from a data source. The POOMA II array facility provides an efficient, high-level interface to engines. POOMA II arrays are declared as follows:

```
Array<2, double, Brick>        A(100, 200);
Array<2, double, Distributed>  B(100, 200, layout);
```

This is a variant of the envelope-letter idiom [4]. Array (the envelope) delegates all operations to the particular sort of engine (the letter) that it contains. However, compile-time polymorphism, rather than run-time polymorphism, is used for faster performance. In POOMA II, the engines own the data and arrays simply provide an interface for viewing and manipulating that data. In this sense they have semantics similar to iterators in the Standard Template Library [5], except that they automatically dereference themselves. To enforce const correctness, POOMA II provides a ConstArray class (similar to the STL const_iterator) that prohibits modification of its elements.

3   Domains and Views

Domain objects represent the region or set of points on which an array will define values. An N-dimensional domain is composed of N one-dimensional domains and represents the tensor product of these domains. POOMA II includes several domain classes:

- Loc<N>: A single point in N-dimensional space.
- Interval<N>: The tensor product of N one-dimensional sequences each having unit stride.
- Range<N>: Similar to Interval<N>, with strides specified at run time.
- Index<N>: Similar to Range<N>, but with special loop-ordering semantics (see below).
- Region<N>: Tensor product of N one-dimensional continuous domains.

Users choose the domain type that best expresses any constraints that they wish to impose on the domain. For example, Interval is used for unit-stride domains and Loc is used for single-point domains. This allows POOMA II to infer properties of the domain at compile time and optimize code accordingly.

One of the primary uses of domains is to specify subsections of Array objects. Subarrays are a common feature of array classes; however, it is often difficult to make such subarrays behave like first-class objects. The POOMA II engine concept provides a clean solution to this problem: subsetting an Array with a domain object creates a new Array that has a *view* engine. For example:

```
Interval<1> I(10);                      // I = {0, 1, ..., 9}
Array<1,double,Brick> A(I);
Range<1> J(0,8,2);                      // J = {0, 2, ..., 8}
Array<1,double,BrickView> B = A(J);
```

The new array B is a view of the even elements of A: {A(0), A(2), ..., A(8)}. Note that views always act as references (i.e., B(0) is an alias for A(0), B(1) is an alias for A(2), etc.). The task of determining the type of view engine to use when subsetting an Array is handled by the NewEngine traits class. Specializations of the class template `NewEngine` define a trait `Type_t` that specifies the type of engine that is created when a particular engine type is subsetted by a particular domain type. Thus, in the above example we could have written:

```
typedef
  NewEngine< Engine<1,double,Brick>, Range<1> >::Type_t View_t;
Array<1,double,View_t::Tag_t> B = A(J);
```

While users can explicitly declare view-engine-based array objects in the manner above, these views will usually be created as temporaries via subscripting and then used in expressions and function calls to specify the elements on which to operate. For example:

```
Interval<1> I(10), I2(2,5);              // I2 = {2, 3, 4, 5}
Array<1,double,Brick> A(I), C(I);
C(I2) = A(I2+1) - A(I2-1);               // C(2) = A(3) - A(1), etc.
```

The final expression builds three temporary views and then executes the expression on these views.

In multidimensional cases, there can be multiple interpretations of certain expressions involving views of arrays. For example, if I and J are domain objects, then what does `A(I,J) = B(J,I)` mean? If I and J are `Interval` objects of equal length, then this would be an element-wise assignment. However, POOMA II's `Index<N>` domain objects have knowledge of their loop ordering. If these domain objects are used, then `A(I,J) = B(J,I)` assigns the transpose of B to A. Thus, the user can choose between tensor-like subscript semantics and Fortran 90 array semantics simply by choosing different domain types.

4   Expressions and Evaluators

Most of the computation in a POOMA II code takes place in mathematical expressions involving several arrays. Expression templates and template metaprograms [7] are used to support an expressive syntax and to implement a number of compile-time optimizations. The most common of these optimizations is converting these high-level expressions into efficient low-level loops.

Expression templates work by storing the parse tree of an expression with operator objects at non-leaf nodes and data objects at the leaves. An expression object is templated on a type that encodes the structure of the parse tree so that the parse tree can be manipulated at compile time to produce efficient code. Consider the sample expression parse tree shown in Fig. 1. PETE encodes this parse tree in an object of type:

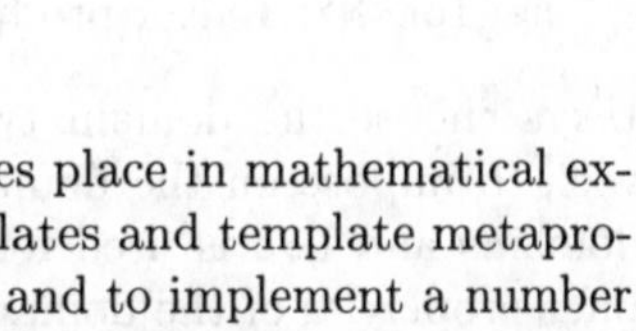

Fig. 1. Parse tree for the expression A = B + 2 * C

```
TBTree< OpAssign, Array1
  TBTree< OpPlus, ConstArray2,
    TBTree< OpMultiply, Scalar<int>, ConstArray3 > > >
```

containing references to arrays A, B and C, and the scalar 2. This expression object can be used to generate an optimized set of loops. However, it does not have array semantics and is not an Array, so it cannot be passed to functions expecting an Array.

The POOMA II engine architecture provides a solution to this problem: the *expression* engine. An expression engine wraps a PETE expression with an engine interface. Values of an expression engine are computed efficiently by the expression-template machinery based on the data referred to in an expression object. With this innovation, the result of an expression involving Array objects is an Array. Thus, users can write functions that operate on expressions by templating them for arbitrary engine types. For example,

```
template<int Dim, class T, class ET>
T trace(const ConstArray<Dim,T,ET> &a) {
   T tr = 0;
   for (int i = 0; i < a.length(0); ++i) { tr += a(i,i); }
   return tr;
}
```

Then trace(B+2*C) sums the diagonal components of B+2*C without computing any of the off-diagonal values.

Expression evaluation is a separate component from the array and expression objects. Evaluators only require a few basic services from arrays and expressions: subsetting, returning an element, getting a domain, etc. Any object that can use those services to evaluate expressions qualifies as an *Evaluator*. Expression evaluation is triggered by the assignment operator of Array, which builds a new Array that has an expression engine and hands it off to an Evaluator. Each expression is defined on a domain, and the Evaluator invokes a function specialized on the type of the domain to evaluate the expression at each point.

For example, suppose an expression is defined on a domain that has only STL-style iterators for looping over the domain. Then, if the domain object is dom and the expression-array object is expr, the inner evaluation loop could look like

```
for (dom::iterator dp = dom.begin(); dp != dom.end(); ++dp)
   expr(*dp);
```

If the domain is a two-dimensional Interval, for which we know that the strides are all unity, the inner loops would look like

```
for (int j = 0; j < dom[1].length(); ++j)
   for (int i = 0; i < dom[0].length(); ++i)
      expr(i, j);
```

The type of inner loop can be determined at compile time since it depends on the type of the domain. That allows the most specialized—and therefore the most efficient—code to be used for the provided data structures.

The Evaluator classes also provide a where-block interface, enabling code such as

```
where(A < 1);
  B = A;
elsewhere();
  B = 1 - A;
endwhere();
```

This code sets array B to A wherever A is less than 1 and 1 - A otherwise. Each call to where(), elsewhere() and endwhere() manipulates state information in the evaluator that influences how expressions are evaluated.

One way to store this state is as a boolean mask array. Because where-blocks can be nested, there must be a stack of such masks, and the top of the stack is the mask for the currently active where-block. Alternatively, one can store a one-dimensional vector of discrete points where the expression is to be evaluated. This would be more efficient than the boolean mask if a small fraction of the mask is true. In either case, the Evaluator extracts an evaluation domain from the where-block expression and evaluates the expression at each point.

The evaluator system is designed to be extensible. Several key extensions are now under development.

Multiblock. Multiblock arrays decompose their data into multiple blocks. The evaluator intersects the subdomains of a multiblock expression, subsets the expression with the intersections, and then evaluates the expression on each subdomain.

SPMD Parallel. In an SPMD parallel environment, the evaluator employs an algorithm such as owner-computes to decide what part of the whole domain should be evaluated on the local processor. It then takes a local view of the expression on that domain. If arrays in the expression have remote data, they must transfer their remote data in order to provide a local view. Once this view is constructed, it can be evaluated efficiently.

Advanced Optimizations. When an expression is ready for evaluation, it need not be evaluated immediately so long as there is a mechanism to account for data dependencies. There are two important reasons for deferred evaluation:

- *Cache optimization.* A given calculation often involves a series of statements that use particular arrays multiple times, but each array is too large to fit in cache. In that case, it is more efficient to block each statement and evaluate one block for a series of statements before working on the next block.
- *Overlapping communication and computation.* Typically the parts of a statement that require communication are along the boundaries of the domain for a given processor. Computation in the interior can proceed while communication needed for the boundaries is taking place.

5 Performance

In order to illustrate performance characteristics of POOMA II, we present a sample results using a stencil benchmark code. A stencil expression is an array expression that involves the same array object evaluated at several nearby points. Such expressions occur frequently in the numerical approximation of partial differential equations, and thus it is important that such expressions be evaluated efficiently.

Consider the simple stencil expression

```
B(I) = K * (A(I+1) - A(I-1))
```

To produce optimal code, the compiler must know that B is not aliased to anything on the right-hand side. It can put &B, &A and K in registers and unroll the loop so that it can save A(I+1) in a register and reuse this value when it needs A(I-1). Accomplishing these optimizations is non-trivial. First, the compiler must be told, via the restrict keyword, that B is not aliased. Second, the compiler must be able to see that both occurrences of A refer to the same array. This is not guaranteed with expression templates, since the pointer to the array being operated on is buried in a TBTree node. Failure to realize this will not only prevent loop unrolling, but also result in the use of extra registers. For large stencils, the compiler may run out of registers ("register spillage"), which greatly impacts performance [6].

These problems can be overcome by encapsulating the stencil operation in a class. A stencil object calculates the value of the stencil given an array and an index. POOMA II stencil objects are fully integrated with the expression-template machinery.

Our stencil benchmark compares four approaches to the evaluation of the 9-point stencil

```
B(I,J) = c * ( A(I-1,J-1) + A(I,J-1) + A(I+1,J-1) +
               A(I-1,J)   + A(I,J)   + A(I+1,J)   +
               A(I-1,J+1) + A(I,J+1) + A(I+1,J+1) );
```

The evaluation methods are C code with restrict, C-style code using POOMA II arrays (C++ indexing), POOMA II code using expression templates (POOMA II Unoptimized), and POOMA II using stencil objects (POOMA II).

The benchmark was performed on an SGI Origin 2000 with 32 KB of primary cache, 4 MB of secondary cache, and a theoretical peak performance of 400 MFlops. Figure 2 shows the results for the four evaluation techniques using $N \times N$ arrays, where N ranges from 10 to 1000. The C code runs significantly faster than the all the C++ versions because it exploits the restrict keyword. For $N > 40$, the arrays are larger than primary cache, but there is little effect on performance. For $N > 400$, the arrays are larger than secondary cache, which leads to a large speed reduction. As the curves for the POOMA II unoptimized and stencil-object versions demonstrate, there is non-zero overhead in the expression-template machinery for small N. The advantage of the stencil-object approach over the unoptimized approach is clearly visible for $N < 100$. This does not persist for large N because the loop is not unrolled (no restrict) and the stencil is not sufficiently large to cause register spillage. An important result is that the stencil-object version performs almost identically to the C++ indexing

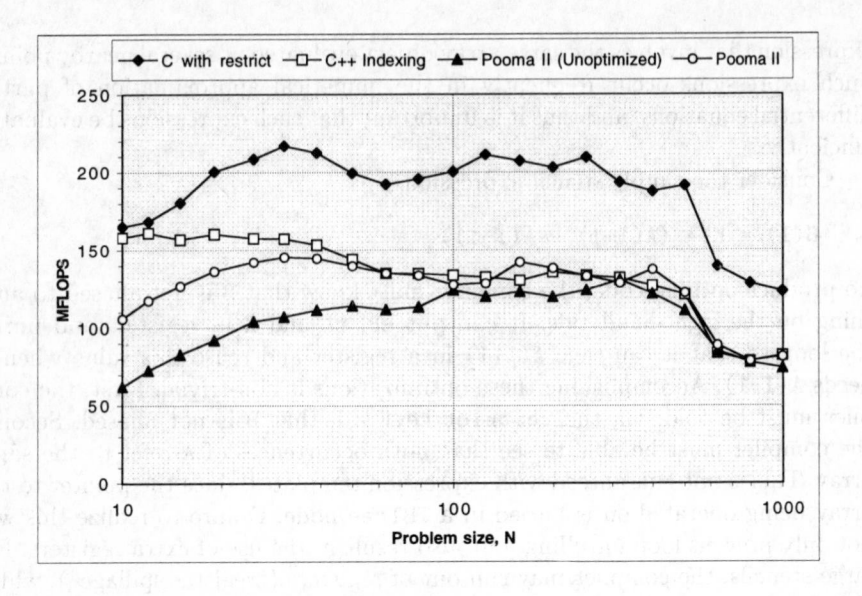

Fig. 2. Stencil benchmark results

version for $N > 30$. Once `restrict` is fully supported for C++, the performance of stencils implemented using POOMA II should closely approach that of C.

References

[1] William Humphrey, Steve Karmesin, Federico Bassetti, and John Reynders. Optimization of data-parallel field expressions in the POOMA framework. In *ISCOPE '97*, December 1997. Marina del Rey, CA.

[2] John Reynders et al. POOMA: A framework for scientific simulations on parallel architectures. In Gregory V. Wilson and Paul Lu, editors, *Parallel Programming using C++*, pages 553–594. MIT Press, 1996.

[3] Todd Veldhuizen. Expression templates. *C++ Report*, June 1995.

[4] James O. Coplien. *Advanced C++ Programming Styles and Idioms*. Addison-Wesley, 1992.

[5] David R. Musser and Atul Saini. *STL Tutorial and Reference Guide: C++ Programming with the Standard Template Library*. Addison-Wesley, 1996.

[6] Federico Bassetti, Kei Davis, and Dan Quinlan. A comparison of performance-enhancing strategies for parallel numerical object-oriented frameworks. In *ISCOPE '97*, December 1997. Marina del Rey, CA.

Author Contacts

Hitoshi Aida, Graduate School of Engineering, University of Tokyo, 7-3-1 Hongo, Bunkyo-ku, Tokyo 113-8656, Japan; aida@sail.t.u-tokyo.ac.jp.

Beat Ammon, Theoretical Physics, ETH Zürich, CH-8093 Zürich, Switzerland; bammon@itp.phys.ethz.ch.

Terumasa Aoki, Research Center for Advanced Science and Technology, University of Tokyo, 4-6-1 Komaba, Meguro-ku, Tokyo 153-8904, Japan; aoki@mpeg.rcast.u-tokyo.ac.jp.

Wolfram T. Arnold, Physics Department, University of Oregon, Eugene, OR 97403; wolfram@darkwing.uoregon.edu.

Kevin Barker, Department of Computer Science and Engineering, University of Notre Dame, Notre Dame, IN 46556; kbarker1@cse.nd.edu.

Federico Bassetti, CIC-19, Los Alamos National Laboratory, Los Alamos, NM 87545; fede@lanl.gov.

Hua Bi, Parallel and Distributed Systems Laboratory, GMD Institute for Computer Architecture and Software Technology, Rudower Chaussee 5, D-12489 Berlin, Germany; bi@first.gmd.de.

Brian Blount, Computer Science Department, University of North Carolina, Chapel Hill, NC 27599-3175; bbount@cs.unc.edu.

Peter A. Buhr, Computer Science Department, University of Waterloo, Waterloo, Ontario, Canada N2L 3G1; pabuhr@uwaterloo.ca.

Siddhartha Chatterjee, Computer Science Department, University of North Carolina, Chapel Hill, NC 27599-3175; sc@cs.unc.edu.

Serge Chaumette, Laboratoire Bordelais de Recherche en Informatique, Université Bordeaux I, 351 Cours de la Libération, 33405 Talence, France; Serge.Chaumette@labri.u-bordeaux.fr.

Malolan Chetlur, Department of ECECS, P.O. Box 210030, University of Cincinnati, Cincinnati, OH 45221-0030; mal@ececs.uc.edu.

Andrew Chien, Computer Science Department, University of Illinois, Urbana, IL 61801; achien@red-herring.cs.uiuc.edu.

Nikos Chrisochoides, Department of Computer Science and Engineering, University of Notre Dame, Notre Dame, IN 46556; nikos@cse.nd.edu.

Timothy Cleland, Advanced Computing Laboratory, Los Alamos National Laboratory, Los Alamos, NM 87545; cleland@lanl.gov.

James Crotinger, CIC-19, Los Alamos National Laboratory, Los Alamos, NM 87545; jac@lanl.gov.

Julian Cummings, Advanced Computing Laboratory, Los Alamos National Laboratory, Los Alamos, NM 87545; julianc@lanl.gov.

Kei Davis, CIC-19, Los Alamos National Laboratory, Los Alamos, NM 87545; kei@lanl.gov.

Robert Denda, Fakultät für Mathematik und Informatik, Universität Mannheim, Mannheim, Germany; rrdenda@uwaterloo.ca.

Florian Dobrian, Computer Science Department, Old Dominion University, Norfolk, VA 23529-0162; dobrian@cs.odu.edu.

Jack Dongarra, Computer Science Department, University of Tennessee, Knoxville, TN 37996-1301; dongarra@cs.utk.edu.

Bishwaroop Ganguly, Computer Science Department, University of Illinois, Urbana, IL 61801; ganguly@red-herring.cs.uiuc.edu.

Al Geist, Oak Ridge National Laboratory, Box 2008, Bldg. 6012, Oak Ridge, TN 37831-6367; gst@ornl.gov.

Jens Gerlach, Tsukuba Research Center of hte Real World Computing Partnership, Tsukuba Mitsui Building, 1-6-1 Takezono, Tsukuba-shi 305-0032, Japan; jens@trc.rwcp.or.jp.

Salman Habib, T-8, Los Alamos National Laboratory, Los Alamos, NM 87545; habib@lanl.gov.

John Hall, CIC-12, Los Alamos National Laboratory, Los Alamos, NM 87545; jxyh@lanl.gov.

Scott Haney, CIC-19, Los Alamos National Laboratory, Los Alamos, NM 87545; swhaney@lanl.gov.

Chris Hawblitzel, Department of Computer Science and Engineering, University of Notre Dame, Notre Dame, IN 46556.

Roger Haydock, Physics Department, University of Oregon, Eugene, OR 97403; haydock@darkwing.uoregon,edu.

Elmar Heeb, Theoretical Physics, ETH Zürich, CH-8093 Zürich, Switzerland; heeb@itp.phys.ethz.ch.

Soichiro Hidaka, Graduate School of Engineering, University of Tokyo, 7-3-1 Hongo, Bunkyo-ku, Tokyo 113-8656, Japan; hidaka@sail.t.u-tokyo.ac.jp.

William Humphrey, Advanced Computing Laboratory, Los Alamos National Laboratory, Los Alamos, NM 87545; bfh@lanl.gov.

Yutaka Ishikawa, Tsukuba Research Center of hte Real World Computing Partnership, Tsukuba Mitsui Building, 1-6-1 Takezono, Tsukuba-shi 305-0032, Japan; ishikawa@trc.rwcp.or.jp.

Laxmikant V. Kalé, Computer Science Department, University of Illinois, Urbana, IL 61801; lkale1@uiuc.edu.

Steve Karmesin, Advanced Computing Laboratory, Los Alamos National Laboratory, Los Alamos, NM 87545; karmesin@lanl.gov.

Matthias Kessler, Parallel and Distributed Systems Laboratory, GMD Institute for Computer Architecture and Software Technology, Rudower Chaussee 5, D-12489 Berlin, Germany; mk@first.gmd.de.

Gary Kumfert, Computer Science Department, Old Dominion University, Norfolk, VA 23529-0162; kumfert@cs.odu.edu.

Andrew Lumsdaine, Department of Computer Science and Engineering, University of Notre Dame, Notre Dame, IN 46556; lums@lsc.nd.edu.

Graham Mark, CIC-12, Los Alamos National Laboratory, Los Alamos, NM 87545; gam@lanl.gov.

Dale E. Martin, Department of ECECS, P.O. Box 210030, University of Cincinnati, Cincinnati, OH 45221-0030; dmartin@ececs.uc.edu.

Paulo Martins, Departamento de Engenharia Informática, Universidade de Coimbra, POLO II, Vila Franca, 3030–Coimbra, Portugal; pmartins@dei.uc.pt.

Mauro Migliardi, Department of Mathematics and Computer Science, Emory University, Atlanta, GA 30322; om@mathcs.emory.edu.

Démian Nave, Department of Computer Science and Engineering, University of Notre Dame, Notre Dame, IN 46556; dnave@cse.nd.edu.

Antonio J. Nebro, Dpto. de Lenguajes y Ciencias de la Computación, Universidad de Málaga, E.T.S.I en Informática, Campus de Teatinos, E29071–Málaga, Spain; antonio@lcc.uma.es.

Sean O'Rourke, CIC-12, Los Alamos National Laboratory, Los Alamos, NM 87545; seano@lanl.gov.

Jim Painter, X-CI, Los Alamos National Laboratory, Los Alamos, NM 87545; jwp@lanl.gov.

Ernesto Pimentel, Dpto. de Lenguajes y Ciencias de la Computación, Universidad de Málaga, E.T.S.I en Informática, Campus de Teatinos, E29071–Málaga, Spain; ernesto@lcc.uma.es.

Alex Pothen, ICASE, NASA Langley Research Center, Hampton, VA 23681-0001; pothen@icase.edu.

Ji Qiang, LANSCE-1, Los Alamos National Laboratory, Los Alamos, NM 87545; jiqiang@lanl.gov.

Dan Quinlan, CIC-19, Los Alamos National Laboratory, Los Alamos, NM 87545; dquinlan@lanl.gov.

Radharamanan Radhakrishnan, Department of ECECS, P.O. Box 210030, University of Cincinnati, Cincinnati, OH 45221-0030; ramanan@ececs.uc.edu.

Jarmo Rantakokko, Department of Scientific Computing, Uppsala University, Box 120, S-751 04, Uppsala, Sweden; jarmo@tdb.uu.se.

Dhananjai Madhava Rao, Department of ECECS, P.O. Box 210030, University of Cincinnati, Cincinnati, OH 45221-0030; dmadhava@ececs.uc.edu.

John Reynders, CIC-DO, Los Alamos National Laboratory, Los Alamos, NM 87545; reynders@lanl.gov.

Robert Ryne, LANSCE-1, Los Alamos National Laboratory, Los Alamos, NM 87545; ryne@lanl.gov.

Tadao Saito, Graduate School of Engineering, University of Tokyo, 7-3-1 Hongo, Bunkyo-ku, Tokyo 113-8656, Japan; saito@sail.t.u-tokyo.ac.jp.

Mitsuhisa Sato, Tsukuba Research Center of hte Real World Computing Partnership, Tsukuba Mitsui Building, 1-6-1 Takezono, Tsukuba-shi 305-0032, Japan; msato@trc.rwcp.or.jp.

Jeremy G. Siek, Department of Computer Science and Engineering, University of Notre Dame, Notre Dame, IN 46556; jsiek@lsc.nd.edu.

João Gabriel Silva, Departamento de Engenharia Informática, Universidade de Coimbra, POLO II, Vila Franca, 3030–Coimbra, Portugal; jgabriel@dei.uc.pt.

Luís Moura Silva, Departamento de Engenharia Informática, Universidade de Coimbra, POLO II, Vila Franca, 3030–Coimbra, Portugal; luis@dei.uc.pt.

Stephen Smith, Advanced Computing Laboratory, Los Alamos National Laboratory, Los Alamos, NM 87545; sa_smith@lanl.gov.

Vaidy Sunderam, Department of Mathematics and Computer Science, Emory University, Atlanta, GA 30322; vss@mathcs.emory.edu.

Masahiro Tatsumi, Nuclear Fuel Industries, Ltd., Software Engineering Division, 950 Ohaza Noda, Kumatori-cho, Sennan-Gun, Osaka 590-0481, Japan; tatsumi@nfi.co.jp.

José Troya, Dpto. de Lenguajes y Ciencias de la Computación, Universidad de Málaga, E.T.S.I en Informática, Campus de Teatinos, E29071–Málaga, Spain; troya@lcc.uma.es.

Matthias Troyer, Theoretical Physics, ETH Zürich, CH-8093 Zürich, Switzerland; troyer@itp.phys.ethz.ch.

Todd L. Veldhuizen, Computer Science Department, Indiana University, Bloomington, IN 47405-4101; tveldhui@extreme.indiana.edu.

Matthias Wilhelmi, Parallel and Distributed Systems Laboratory, GMD Institute for Computer Architecture and Software Technology, Rudower Chaussee 5, D-12489 Berlin, Germany; wilhelmi@first.gmd.de.

Timothy J. Williams, CIC-19, Los Alamos National Laboratory, Los Alamos, NM 87545; zippy@lanl.gov.

Philip A. Wilsey, Department of ECECS, P.O. Box 210030, University of Cincinnati, Cincinnati, OH 45221-0030; phil.wilsey@uc.edu.

Akio Yamamoto, Nuclear Fuel Industries, Ltd., Software Engineering Division, 950 Ohaza Noda, Kumatori-cho, Sennan-Gun, Osaka 590-0481, Japan; a-yama@nfi.co.jp.

Josh Yelon, Computer Science Department, University of Illinois, Urbana, IL 61801; jyelon@cs.uiuc.edu.

Mark Zander, CIC-12, Los Alamos National Laboratory, Los Alamos, NM 87545; zander@lanl.gov.

Author Index

Aida, H., 151
Ammon, B., 191
Aoki, T., 151
Arnold, W., 199

Barker, K., 71
Bassetti, F., 107
Bi, H., 175
Blount, B., 35
Buhr, P., 159

Chatterjee, S., 35
Chaumette, S., 135
Chetlur, M., 13
Chien, A., 47
Chrisochoides, N., 71
Cleland, T., 25
Crotinger, J., 231
Cummings, J., 25, 231

Davis, K., 107
Denda, R., 159
Dobrian, F., 207
Dongarra, J., 127

Ganguly, B., 47
Geist, A., 127
Gerlach, J., 215

Habib, S., 25
Hall, J., 183
Haney, S., 231
Hawblitzel, C., 71
Haydock, R., 199
Heeb, E., 191
Hidaka, S., 151
Humphrey, W., 25, 231

Ishikawa, Y., 215

Kalé, L., 143
Karmesin, S., 231
Kessler, M., 175
Kumfert, G., 95, 207

Lumsdaine, A., 59

Mark, G., 25
Martin, D., 13
Martins, P., 119
Migliardi, M., 127

Nave, D., 71
Nebro, A., 167

O'Rourke, S., 183

Painter, J., 183
Pimentel, E., 167
Pothen, A., 95, 207

Qiang, J., 25
Quinlan, D., 107

Radhakrishnan, R., 13
Rantakokko, J., 83
Rao, D., 13
Reynders, J., 231
Ryne, R., 25

Saito, T., 151
Sato, M., 215
Siek, J., 59
Silva, J., 119
Silva, L., 119
Smith, S., 231
Sunderam, V., 127

Tatsumi, M., 1
Troya, J., 167
Troyer, M., 191

Veldhuizen, T., 223

Wilhelmi, M., 175
Williams, T., 231
Wilsey, P., 13

Yamamoto, A., 1
Yelon, J., 143

Zander, M., 183

Springer
and the
environment

At Springer we firmly believe that an international science publisher has a special obligation to the environment, and our corporate policies consistently reflect this conviction.

We also expect our business partners – paper mills, printers, packaging manufacturers, etc. – to commit themselves to using materials and production processes that do not harm the environment. The paper in this book is made from low- or no-chlorine pulp and is acid free, in conformance with international standards for paper permanency.

Vol. 1490: C. Palamidessi, H. Glaser, K. Meinke (Eds.), Principles of Declarative Programming. Proceedings, 1998. XI, 497 pages. 1998.

Vol. 1491: W. Reisig, G. Rozenberg (Eds.), Lectures on Petri Nets I: Basic Models. XII, 683 pages. 1998.

Vol. 1492: W. Reisig, G. Rozenberg (Eds.), Lectures on Petri Nets II: Applications. XII, 479 pages. 1998.

Vol. 1493: J.P. Bowen, A. Fett, M.G. Hinchey (Eds.), ZUM '98: The Z Formal Specification Notation. Proceedings, 1998. XV, 417 pages. 1998.

Vol. 1494: G. Rozenberg, F. Vaandrager (Eds.), Lectures on Embedded Systems. Proceedings, 1996. VIII, 423 pages. 1998.

Vol. 1495: T. Andreasen, H. Christiansen, H.L. Larsen (Eds.), Flexible Query Answering Systems. IX, 393 pages. 1998. (Subseries LNAI).

Vol. 1496: W.M. Wells, A. Colchester, S. Delp (Eds.), Medical Image Computing and Computer-Assisted Intervention – MICCAI'98. Proceedings, 1998. XXII, 1256 pages. 1998.

Vol. 1497: V. Alexandrov, J. Dongarra (Eds.), Recent Advances in Parallel Virtual Machine and Message Passing Interface. Proceedings, 1998. XII, 412 pages. 1998.

Vol. 1498: A.E. Eiben, T. Bäck, M. Schoenauer, H.-P. Schwefel (Eds.), Parallel Problem Solving from Nature – PPSN V. Proceedings, 1998. XXIII, 1041 pages. 1998.

Vol. 1499: S. Kutten (Ed.), Distributed Computing. Proceedings, 1998. XII, 419 pages. 1998.

Vol. 1501: M.M. Richter, C.H. Smith, R. Wiehagen, T. Zeugmann (Eds.), Algorithmic Learning Theory. Proceedings, 1998. XI, 439 pages. 1998. (Subseries LNAI).

Vol. 1502: G. Antoniou, J. Slaney (Eds.), Advanced Topics in Artificial Intelligence. Proceedings, 1998. XI, 333 pages. 1998. (Subseries LNAI).

Vol. 1503: G. Levi (Ed.), Static Analysis. Proceedings, 1998. IX, 383 pages. 1998.

Vol. 1504: O. Herzog, A. Günter (Eds.), KI-98: Advances in Artificial Intelligence. Proceedings, 1998. XI, 355 pages. 1998. (Subseries LNAI).

Vol. 1505: D. Caromel, R.R. Oldehoeft, M. Tholburn (Eds.), Computing in Object-Oriented Parallel Environments. Proceedings, 1998. XI, 243 pages. 1998.

Vol. 1506: R. Koch, L. Van Gool (Eds.), 3D Structure from Multiple Images of Large-Scale Environments. Proceedings, 1998. VIII, 347 pages. 1998.

Vol. 1507: T.W. Ling, S. Ram, M.L. Lee (Eds.), Conceptual Modeling – ER '98. Proceedings, 1998. XVI, 482 pages. 1998.

Vol. 1508: S. Jajodia, M.T. Özsu, A. Dogac (Eds.), Advances in Multimedia Information Systems. Proceedings, 1998. VIII, 207 pages. 1998.

Vol. 1510: J.M. Zytkow, M. Quafafou (Eds.), Principles of Data Mining and Knowledge Discovery. Proceedings, 1998. XI, 482 pages. 1998. (Subseries LNAI).

Vol. 1511: D. O'Hallaron (Ed.), Languages, Compilers, and Run-Time Systems for Scalable Computers. Proceedings, 1998. IX, 412 pages. 1998.

Vol. 1512: E. Giménez, C. Paulin-Mohring (Eds.), Types for Proofs and Programs. Proceedings, 1996. VIII, 373 pages. 1998.

Vol. 1513: C. Nikolaou, C. Stephanidis (Eds.), Research and Advanced Technology for Digital Libraries. Proceedings, 1998. XV, 912 pages. 1998.

Vol. 1514: K. Ohta, D. Pei (Eds.), Advances in Cryptology – ASIACRYPT'98. Proceedings, 1998. XII, 436 pages. 1998.

Vol. 1515: F. Moreira de Oliveira (Ed.), Advances in Artificial Intelligence. Proceedings, 1998. X, 259 pages. 1998. (Subseries LNAI).

Vol. 1516: W. Ehrenberger (Ed.), Computer Safety, Reliability and Security. Proceedings, 1998. XVI, 392 pages. 1998.

Vol. 1517: J. Hromkovič, O. Sýkora (Eds.), Graph-Theoretic Concepts in Computer Science. Proceedings, 1998. X, 385 pages. 1998.

Vol. 1518: M. Luby, J. Rolim, M. Serna (Eds.), Randomization and Approximation Techniques in Computer Science. Proceedings, 1998. IX, 385 pages. 1998.

Vol. 1520: M. Maher, J.-F. Puget (Eds.), Principles and Practice of Constraint Programming - CP98. Proceedings, 1998. XI, 482 pages. 1998.

Vol. 1521: B. Rovan (Ed.), SOFSEM'98: Theory and Practice of Informatics. Proceedings, 1998. XI, 453 pages. 1998.

Vol. 1522: G. Gopalakrishnan, P. Windley (Eds.), Formal Methods in Computer-Aided Design. Proceedings, 1998. IX, 529 pages. 1998.

Vol. 1524: G.B. Orr, K.-R. Müller (Eds.), Neural Networks: Tricks of the Trade. VI, 432 pages. 1998.

Vol. 1525: D. Aucsmith (Ed.), Information Hiding. Proceedings, 1998. IX, 369 pages. 1998.

Vol. 1526: M. Broy, B. Rumpe (Eds.), Requirements Targeting Software and Systems Engineering. Proceedings, 1997. VIII, 357 pages. 1998.

Vol. 1529: D. Farwell, L. Gerber, E. Hovy (Eds.), Machine Translation and the Information Soup. Proceedings, 1998. XIX, 532 pages. 1998. (Subseries LNAI).

Vol. 1530: V. Arvind, R. Ramanujam (Eds.), Foundations of Software Technology and Theoretical Computer Science. XII, 369 pages. 1998.

Vol. 1531: H.-Y. Lee, H. Motoda (Eds.), PRICAI'98: Topics in Artificial Intelligence. XIX, 646 pages. 1998. (Subseries LNAI).

Vol. 1096: T. Schael, Workflow Management Systems for Process Organisations. Second Edition. XII, 229 pages. 1998.

Vol. 1532: S. Arikawa, H. Motoda (Eds.), Discovery Science. Proceedings, 1998. XI, 456 pages. 1998. (Subseries LNAI).

Vol. 1533: K.-Y. Chwa, O.H. Ibarra (Eds.), Algorithms and Computation. Proceedings, 1998. XIII, 478 pages. 1998.

Vol. 1538: J. Hsiang, A. Ohori (Eds.), Advances in Computing Science – ASIAN'98. Proceedings, 1998. X, 305 pages. 1998.

Lecture Notes in Computer Science

For information about Vols. 1–1454
please contact your bookseller or Springer-Verlag

Vol. 1455: A. Hunter, S. Parsons (Eds.), Applications of Uncertainty Formalisms. VIII, 474 pages. 1998. (Subseries LNAI).

Vol. 1456: A. Drogoul, M. Tambe, T. Fukuda (Eds.), Collective Robotics. Proceedings, 1998. VII, 161 pages. 1998. (Subseries LNAI).

Vol. 1457: A. Ferreira, J. Rolim, H. Simon, S.-H. Teng (Eds.), Solving Irregularly Structured Problems in Prallel. Proceedings, 1998. X, 408 pages. 1998.

Vol. 1458: V.O. Mittal, H.A. Yanco, J. Aronis, R-. Simpson (Eds.), Assistive Technology in Artificial Intelligence. X, 273 pages. 1998. (Subseries LNAI).

Vol. 1459: D.G. Feitelson, L. Rudolph (Eds.), Job Scheduling Strategies for Parallel Processing. Proceedings, 1998. VII, 257 pages. 1998.

Vol. 1460: G. Quirchmayr, E. Schweighofer, T.J.M. Bench-Capon (Eds.), Database and Expert Systems Applications. Proceedings, 1998. XVI, 905 pages. 1998.

Vol. 1461: G. Bilardi, G.F. Italiano, A. Pietracaprina, G. Pucci (Eds.), Algorithms – ESA'98. Proceedings, 1998. XII, 516 pages. 1998.

Vol. 1462: H. Krawczyk (Ed.), Advances in Cryptology - CRYPTO '98. Proceedings, 1998. XII, 519 pages. 1998.

Vol. 1463: N.E. Fuchs (Ed.), Logic Program Synthesis and Transformation. Proceedings, 1997. X, 343 pages. 1998.

Vol. 1464: H.H.S. Ip, A.W.M. Smeulders (Eds.), Multimedia Information Analysis and Retrieval. Proceedings, 1998. VIII, 264 pages. 1998.

Vol. 1465: R. Hirschfeld (Ed.), Financial Cryptography. Proceedings, 1998. VIII, 311 pages. 1998.

Vol. 1466: D. Sangiorgi, R. de Simone (Eds.), CONCUR'98: Concurrency Theory. Proceedings, 1998. XI, 657 pages. 1998.

Vol. 1467: C. Clack, K. Hammond, T. Davie (Eds.), Implementation of Functional Languages. Proceedings, 1997. X, 375 pages. 1998.

Vol. 1468: P. Husbands, J.-A. Meyer (Eds.), Evolutionary Robotics. Proceedings, 1998. VIII, 247 pages. 1998.

Vol. 1469: R. Puigjaner, N.N. Savino, B. Serra (Eds.), Computer Performance Evaluation. Proceedings, 1998. XIII, 376 pages. 1998.

Vol. 1470: D. Pritchard, J. Reeve (Eds.), Euro-Par'98: Parallel Processing. Proceedings, 1998. XXII, 1157 pages. 1998.

Vol. 1471: J. Dix, L. Moniz Pereira, T.C. Przymusinski (Eds.), Logic Programming and Knowledge Representation. Proceedings, 1997. IX, 246 pages. 1998. (Subseries LNAI).

Vol. 1472: B. Freitag, H. Decker, M. Kifer, A. Voronkov (Eds.), Transactions and Change in Logic Databases. Proceedings, 1996, 1997. X, 396 pages. 1998.

Vol. 1473: X. Leroy, A. Ohori (Eds.), Types in Compilation. Proceedings, 1998. VIII, 299 pages. 1998.

Vol. 1474: F. Mueller, A. Bestavros (Eds.), Languages, Compilers, and Tools for Embedded Systems. Proceedings, 1998. XIV, 261 pages. 1998.

Vol. 1475: W. Litwin, T. Morzy, G. Vossen (Eds.), Advances in Databases and Information Systems. Proceedings, 1998. XIV, 369 pages. 1998.

Vol. 1476: J. Calmet, J. Plaza (Eds.), Artificial Intelligence and Symbolic Computation. Proceedings, 1998. XI, 309 pages. 1998. (Subseries LNAI).

Vol. 1477: K. Rothermel, F. Hohl (Eds.), Mobile Agents. Proceedings, 1998. VIII, 285 pages. 1998.

Vol. 1478: M. Sipper, D. Mange, A. Pérez-Uribe (Eds.), Evolvable Systems: From Biology to Hardware. Proceedings, 1998. IX, 382 pages. 1998.

Vol. 1479: J. Grundy, M. Newey (Eds.), Theorem Proving in Higher Order Logics. Proceedings, 1998. VIII, 497 pages. 1998.

Vol. 1480: F. Giunchiglia (Ed.), Artificial Intelligence: Methodology, Systems, and Applications. Proceedings, 1998. IX, 502 pages. 1998. (Subseries LNAI).

Vol. 1481: E.V. Munson, C. Nicholas, D. Wood (Eds.), Principles of Digital Document Processing. Proceedings, 1998. VII, 152 pages. 1998.

Vol. 1482: R.W. Hartenstein, A. Keevallik (Eds.), Field-Programmable Logic and Applications. Proceedings, 1998. XI, 533 pages. 1998.

Vol. 1483: T. Plagemann, V. Goebel (Eds.), Interactive Distributed Multimedia Systems and Telecommunication Services. Proceedings, 1998. XV, 326 pages. 1998.

Vol. 1484: H. Coelho (Ed.), Progress in Artificial Intelligence – IBERAMIA 98. Proceedings, 1998. XIII, 421 pages. 1998. (Subseries LNAI).

Vol. 1485: J.-J. Quisquater, Y. Deswarte, C. Meadows, D. Gollmann (Eds.), Computer Security – ESORICS 98. Proceedings, 1998. X, 377 pages. 1998.

Vol. 1486: A.P. Ravn, H. Rischel (Eds.), Formal Techniques in Real-Time and Fault-Tolerant Systems. Proceedings, 1998. VIII, 339 pages. 1998.

Vol. 1487: V. Gruhn (Ed.), Software Process Technology. Proceedings, 1998. VIII, 157 pages. 1998.

Vol. 1488: B. Smyth, P. Cunningham (Eds.), Advances in Case-Based Reasoning. Proceedings, 1998. XI, 482 pages. 1998. (Subseries LNAI).

Vol. 1489: J. Dix, L. Fariñas del Cerro, U. Furbach (Eds.), Logics in Artificial Intelligence. Proceedings, 1998. X, 391 pages. 1998. (Subseries LNAI).